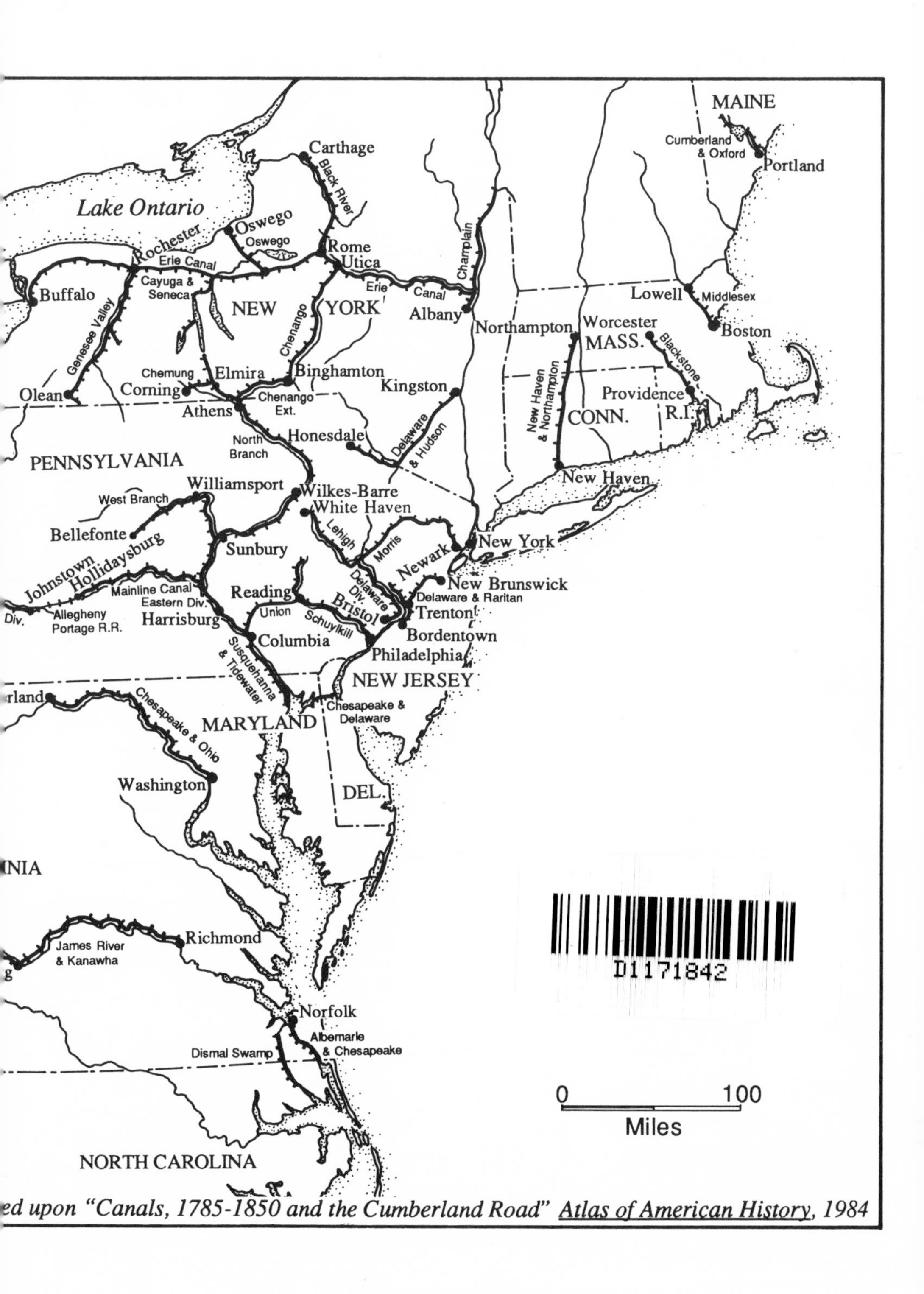

...ed upon "Canals, 1785-1850 and the Cumberland Road" *Atlas of American History*, *1984*

Canals for a Nation

CANALS FOR A NATION

THE Canal Era IN THE United States 1790-1860

RONALD E. SHAW

THE UNIVERSITY PRESS OF KENTUCKY

Scholarly publisher for the Commonwealth,
serving Bellarmine College, Berea College, Centre
College of Kentucky, Eastern Kentucky University,
The Filson Club, Georgetown College, Kentucky
Historical Society, Kentucky State University,
Morehead State University, Murray State University
Northern Kentucky University, Transylvania University,
University of Kentucky, University of Louisville,
and Western Kentucky University.

Editorial and Sales Offices: Lexington, Kentucky 40508-4008

Library of Congress Cataloging-in-Publication Data

Shaw, Ronald E.
Canals for a nation : the canal era in the United States,
1790-1860 / Ronald E. Shaw.
p. cm.
Includes bibliographical references and index.
ISBN 0-8131-1701-1 (acid-free paper)
1. Canals—United States—History. 2. Inland navigation—United
States—History. I. Title.
TC623.S53 1990
386'.4'0973—dc20 90-42008
CIP

This book is printed on acid-free paper meeting
the requirements of the American National Standard
for Permanence of Paper for Printed Library Materials. ♾

For Judy

Contents

Preface

The Canal Era was a major phase of America's nineteenth century transportation revolution. Canals lowered transportation costs, carried a vast grain trade from western farms to eastern ports, and delivered Pennsylvania coal to New Jersey and New York. They created new towns and cities and contributed to American economic growth.

My earlier study of the Erie Canal led me to undertake a survey of the history of American canals, which would reflect the economic studies that have emphasized the role of government and mixed enterprise in canal building, as well as the more recent emphasis on canals and the preservation of republicanism in the new American nation. In this synthesis my approach is comparative, showing the transfer of European technology to America, the remarkable success of the Erie Canal, and the competition for trunk-line routes linking eastern cities to the trans-Appalachian West.

I have found common themes in the work of engineers who carried their skills from state to state, the near-heroic figures who devoted their lives to canals; similar crises in canal financing; the need to create and sustain political support; the mixed enterprise underlying both publicly and privately-built canals; the presence almost everywhere of Irish laborers enduring brutal conditions of work; and the heady response to canal travel.

For me, American canals were audacious achievements of engineering and construction, often in nearly impossible terrain. They were incremental triumphs accomplished in spite of all the uncertainties of the political process. Localism and self-serving political logrolling were clearly evident, but I have attempted to describe American canals as a transportation network that was frequently justified or celebrated in expressions of nationalism. One aspect of this nationalism was the preservation of republicanism, to be strengthened by canals that would bind the union together. This is not a quantitative study, but it does reveal the great economic impact of American canals, whether or not they were directly profitable or were more developmental in influence. My background has been that of an academic historian, but I have also incorporated some of the work of the burgeoning canal societies, which should be recognized for their expertise in interpreting and preserving the physical remains of the canals.

Research for this study has been supported by Miami University through summer research fellowships, assigned research leaves, and research grants. The Indiana Historical Society has kindly granted permission to use in chapters 5, 6, and 7 materials drawn from my chapter, "The Canal Era in the Old Northwest," in *Transportation and the Early Nation* (Indianapolis 1982).

I am indebted to many people for their contributions to this project. Ralph D. Gray assisted me on recent canal scholarship and used his careful editorial hand as he read an early version of the manuscript. Lance E. Metz reviewed my initial manuscript and suggested revisions, contributing his special knowledge of the Pennsylvania canals. The support of Wm. Jerome Crouch, whose editing polished my book on the Erie Canal and whose patience helped to sustain the interest of the University Press of Kentucky in the present volume, has been especially appreciated. David M. Fahey has been a friend in the truest sense of the word and has made timely suggestions to keep this project on track toward completion. At the Ohio Historical Society, Steven C. Gordon provided materials for maps and illustrations.

I wish to acknowledge the assistance of the reference staff at King Library of Miami University, especially of C. Martin Miller, Jenny Presnell, and William Wortman. In the harried time of proofreading and indexing I have had the assistance of William R. Wantland, a doctoral student in history at Miami University.

Again, I owe a special debt to my wife, Judith M. Shaw, for her continuing support and perceptive criticisms, which have made this work a shared endeavor.

1

Pioneer Canals and Republican Improvements

Like so much in American history, the first canals were derivative from those of Europe. The Dutch, the French, and, especially in the latter half of the eighteenth century, the English carried the technology of canal building forward from its ancient origins. The Dutch developed a canal-integrated economy; the French completed the great Languedoc Canal (Canal du Midi), 148 miles long, in 1681; and the English midlands were crisscrossed by a rapidly expanding network of narrow canals from Regents Park in London north to Scotland. Many of the English canals were being built just as the Canal Era opened in the United States.[1]

In America the Canal Era emerged out of the transfer of European technology, stimulated by the ideal of a water-connected society in a garden environment.[2] English examples in canal building were followed most frequently, and the English engineers William Weston and Benjamin Latrobe did the surveys for most of the pioneer canals in the 1790s. Weston brought the Troughton Y level, a spirit level attached to a telescope, which was probably the first leveling instrument in America.[3] American engineers such as Loammi Baldwin, Jr., Canvass White, William Strickland, and Robert Mills, who lacked formal training, tramped the English canals and above all sought English experience in their crucial quest for a workable underwater cement.

The engineers who struggled to build the first American canals faced the most elementary yet trouble-plagued problems: how to take an accurate level, often with only the equipment of a country surveyor; how to dig a canal channel and remove the earth most efficiently; how best to cut through tree roots; how to use blasting powder to remove embedded rock; how to keep canal banks from leaking by "puddling"; how to mix a permanent underwater cement, to design lock gates, to make valves for the gradual release of water, and to create locks and gates that could be operated easily by hand power.

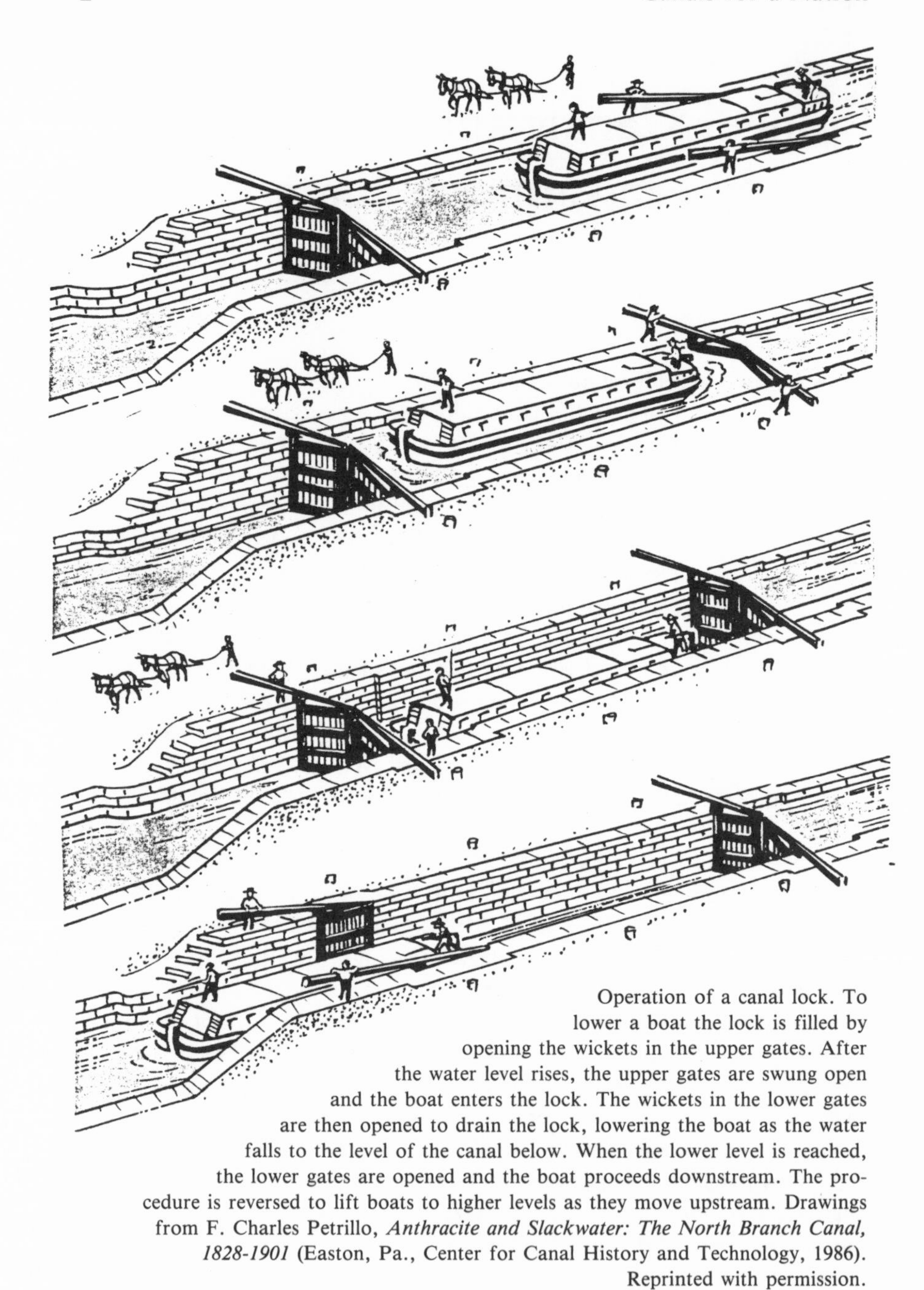

Operation of a canal lock. To lower a boat the lock is filled by opening the wickets in the upper gates. After the water level rises, the upper gates are swung open and the boat enters the lock. The wickets in the lower gates are then opened to drain the lock, lowering the boat as the water falls to the level of the canal below. When the lower level is reached, the lower gates are opened and the boat proceeds downstream. The procedure is reversed to lift boats to higher levels as they move upstream. Drawings from F. Charles Petrillo, *Anthracite and Slackwater: The North Branch Canal, 1828-1901* (Easton, Pa., Center for Canal History and Technology, 1986). Reprinted with permission.

The first canals of the 1790s were tiny in size, painfully slow in construction, and financed by companies that were always short of funds. Yet these companies were led by the landed or commercial gentry of the time, with names distinguished for political leadership or social standing. These early canal builders were imbued with a spirit of improvement that has often been obscured by their political contributions in creating a new nation. Although the canals they built were short waterways and their companies were small, they were attempting to achieve the same republican ends that they pursued in better-known positions of public service.[4]

River improvements in the 1790s and early 1800s were too numerous to trace, but several short canals stand out in the pioneer experiences of canal building in America. The Schuylkill and Susquehanna Canal in Pennsylvania and the Potomac Canal in Virginia started toward a common destination in the Ohio Valley. Navigation of the Susquehanna River was improved by the Conewago and Susquehanna canals. Along the Atlantic coast, the Dismal Swamp Canal connected Chesapeake Bay to Albermarle Sound. In the northeastern states, the Western Inland Lock Navigation Company built important canals in the Mohawk Valley; and in Massachusetts, the Middlesex Canal reached from Boston up to the Merrimack River and New Hampshire. Far to the south the Santee and Cooper Canal brought the trade of the upland rivers to Charleston. Not only did canal ventures in Pennsylvania and Virginia share a common goal in the improvement of water transportation to the Ohio Valley, but their early engineering experiences contributed to the building of other pioneer canals in the 1790s.[5]

In Pennsylvania, long before the Revolution, a group of Philadelphians including Thomas Gilpin, David Rittenhouse, Benjamin Franklin, William Smith, and Thomas Mifflin began to make plans for water improvements from Philadelphia to the Susquehanna Valley. They developed what Darwin H. Stapleton has called the "Philadelphia plan," which considered three routes by which Philadelphia might gain the trade of the Susquehanna Valley, one by a canal across the Delmarva Peninsula south of Philadelphia, one by a road from the Susquehanna to a river port near Philadelphia, and a third from the Schuylkill River to the Susquehanna by a canal along their tributaries, the Tulpehocken and the Swatara rivers. The last of these three routes became the project for the Schuylkill and Susquehanna Canal.[6]

The building of this canal became the goal of the Society for the

Improvement of Roads and Inland Navigation in 1789, organized by Robert Morris, Rittenhouse, Smith, and the Philadelphia land speculator John Nicholson. The society's president was Morris, who signed its memorial to the Pennsylvania Assembly, marking out a water route to the interior. The canal would go up the Schuylkill and Tulpehocken rivers and cross the summit level near Lebanon by a canal, to reach the westward-flowing Quitapahilla River and the Swatara, which was tributary to the Susquehanna.

In response to the society's proposal and the Morris memorial, the Pennsylvania legislature incorporated the Schuylkill and Susquehanna Navigation Company in 1791 to build the Schuylkill and Susquehanna Canal and improve the navigation of the Schuylkill River. To connect this canal to Philadelphia, the Delaware and Schuylkill Navigation Company was chartered to build another canal to run from Norristown on the Schuylkill for seventeen miles to the Delaware River near Philadelphia. Robert Morris was president of both companies.

The Inland Navigation Society and the Morris memorial projected a route to continue west through the Susquehanna Valley to the Juniata River and up that valley to Poplar Run. There a short mountainous portage was necessary to reach the Conemaugh, the Kiskiminitas, and the Allegheny rivers, which led finally to Pittsburgh on the Ohio River. The total distance of this long water route was calculated to be 426 miles, and Morris and his associates concluded that it would allow goods to be carried more cheaply from Philadelphia to the Ohio River than from any other Atlantic port. This great route, Governor Mifflin told the legislature in 1790, was "a natural avenue from the shores of the Atlantic to the vast regions of the western territory," connecting "the extreme members of the union," and offering rewards beyond imagination.[7]

Almost desperate for professional engineering skill, the Schuylkill and Susquehanna Canal Company brought the English engineer William Weston to supervise construction on the Schuylkill and Susquehanna Canal. Weston was probably the son of Samuel Weston, an experienced English canal engineer, and he was recommended by William Jessop, who was then the leading engineer in England. William Weston had trained under James Brindley and had worked on canals in Ireland and central England. He was offered the then stunning stipend of £800 a year to work for the company for five years. When he arrived in Pennsylvania in 1793, he found more than six hundred men at work at Norristown and on the summit level on the canal between Lebanon and Myerstown.

quehanna Valley were brought by wagon from Philadelphia to enter the river at Columbia. Although Latrobe did improve some rivers in Pennsylvania, the state was so anxious to protect the trade over the Philadelphia and Lancaster Turnpike that it refused to cooperate with Maryland in improving the river above Maryland's Susquehanna Canal. The canal was sold in 1817 at a loss to the proprietors.[13]

At the same time, Latrobe began work on the Chesapeake and Delaware Canal, which was tied to the lower Susquehanna improvements and begun by a company chartered in Pennsylvania, Maryland, and Delaware. He began work on this project to connect the Chesapeake and Delaware bays in 1803 and continued until funds were exhausted in 1805 and work stopped. On this canal he trained two future canal engineers, William Strickland and Robert Mills.

While Pennsylvanians sought to improve the Schuylkill and Susquehanna routes to the Ohio Valley, Virginians sought to improve the Potomac route to the West, believing it superior to any other. The guiding spirit in Virginia's effort was George Washington, who, after the Revolutionary War, turned his attention to canals. He made a tour of the waterways of eastern New York in 1783 and the next year undertook a great circular expedition up the Potomac, across the mountains to the Ohio River and Lake Erie, and back by the Kanawha and James rivers. His interest in the improvement of the Potomac had grown following his expedition to Fort Dusquesne in 1784, and his belief in canals was strengthened by his long experience as a surveyor.

When the Potomac Company was organized in 1785 to improve the Potomac navigation, Washington became its president. He believed such improvements would prove profitable to his land speculations, would be critical to the interests of his state of Virginia, and would help to preserve the American Union. Washington worried that western settlers would become commercially dependent upon the British and the Spanish "at the flanks and rear" of the United States. On his return from his long western journey in 1784, he wrote in his diary, "The Western Settlers—from my own observation—stand as it were upon a pivet—the touch of a feather would almost incline them any way." A few days later he wrote almost the same words in a letter to Virginia's governor, Benjamin Harrison.[14]

Washington estimated that the improvement of sixty miles of Potomac River navigation from Georgetown to Harpers Ferry could ultimately replace two hundred miles of land travel to the Ohio River. This could be done by improving the Potomac, adding a portage road from the Cumberland to the Monongahela River, and following the Monongahela to the Ohio. In a larger scheme he wished to turn the trade of the interior to the Potomac and his native Virginia, rather than allowing it to be drawn to the Mohawk Valley, the Hudson River, and New York. Since the Potomac is in Maryland, his efforts required the cooperation of that state, which he achieved in the Mount Vernon Compact of 1785, in spite of the resistance of Baltimore merchants.

Both James Madison and Thomas Jefferson supported the Potomac Company. Madison saw to the company's interests in the Virginia legislature. Jefferson wrote to Washington in 1784 that the Potomac offered the shortest route to the "Western world." In the "rivalship between the Hudson and Patowmac" he found the Potomac route 730 miles nearer from the Ohio River to Alexandria than to New York. "Nature then has declared in favour of the Potowmack," he added, "and . . . it behoves [sic] us then to open our doors to it."[15]

On the Potomac River there were five rapids or falls to be passed. Ten miles above Georgetown the Potomac Company built a canal with five locks around the largest obstruction, the Great Falls. James Rumsey, the chief engineer, was really a surveyor-mechanic, better known for his later attempts at steam navigation. William Weston came to the Great Falls works in 1795 and gave his advice. Washington continued to draw upon Weston by letters, and Weston returned briefly in 1796. Labor was supplied by free, indentured, and slave workers. Stock subscriptions given by merchants and planters from Alexandria and Georgetown were paid on a "pledge now–pay later" basis, and land speculations such as those of Richard Henry Lee and James Madison at Great Falls became enmeshed with canal stock pledges, to the detriment of the latter. Only an additional stock subscription by the Maryland legislature allowed the company to open the locks at Great Falls in 1802. Through them passed the little boats that entered the Potomac at Lock's Cove, from which they were poled to Georgetown, Washington, or Alexandria.

By then the company had abandoned its plans for connections to the Ohio and limited its work to the improvement of the Potomac. But traffic on the Potomac was limited by lack of the anticipated settlement upriver and by changes in the European market so that by 1822 the company had spent $729,387 and had debts of $175,886.[16] Still, the Potomac Company's improvements served the Potomac trade and kept alive the ultimate goal of a canal to the Ohio River. Its faltering finances demonstrated the need for the national aid that would make possible its successor, the Chesapeake and Ohio Canal, which would pass beside the Potomac to Harpers Ferry by 1830.

In 1785, the year the Potomac Company was chartered, George Washington also sponsored the James River Company farther south in Virginia. Under the active management of Edmund Randolph, this company built a seven-mile canal from Richmond to the falls at Westham, which began to operate in 1795. And to the southwest, the Dismal Swamp Canal was built to connect Chesapeake Bay and the Pasquotank River, which led to Albermarle Sound off North Carolina. This waterway, opened in 1794 to small boats and enlarged in 1807 to take six-foot-wide flatboats, was fully opened in 1812. It had no locks, but it was one of the longest of the pioneer canals, thirty-two feet wide and twenty-two miles long.[17]

In the Northeast, small canals were built during the 1790s in Connecticut and New York. On the Connecticut River, an essential route for settlement in western New England, a two-and-a-half-mile canal at South Hadley Falls was built between 1792 and 1795. This little waterway boasted the ingenious South Hadley Inclined Plane, designed by Benjamin Prescott. Powered by two water wheels, it carried loaded boats on a carriage from one level to another and remained in service for more than half a century. A canal to connect the Connecticut River above Hartford with Boston was proposed by Secretary of War Henry Knox in 1790. Knox brought a surveyor from England, Captain John Hills, who examined a route but took no levels, and the company chartered for the canal was never organized. But across the Hudson River in New York the Western Inland Lock Navigation Company, headed by General Philip Schuyler, made a significant effort to improve the navigation of the Mohawk River.

Schuyler was typical of the public-spirited, land-speculating investors who plunged into canal building to improve water transportation to the West. Dixon Ryan Fox has described him as "the chief patron of internal improvements" in New York.[18] Working with Schuyler was the more visionary Elkanah Watson, who visited George Washington at Mount Vernon and called himself Washington's "canal disciple." The Western Inland Lock Navigation Company was incorporated in 1792 and worked for years to build a portage canal around the Little Falls of the Mohawk and make other improvements to open navigation between the

Hudson and Lakes Seneca and Ontario. Schuyler, Watson, and others also organized the Northern Inland Lock Navigation Company to improve navigation between the Hudson and Lake Champlain.[19]

Unable to find a competent engineer, Schuyler began to superintend canal operations himself, without, as he wrote, "the least practical experience in the business." Schuyler had seen the new canals in England when he visited there in 1761, and in 1793 he wrote to Robert Morris asking for the loan of the English engineer William Weston. Weston arrived in 1795, examined the work, and left instructions for the company to carry out. A year later, Weston returned and took charge of the work. Among his assistants was Benjamin Wright, who was learning the profession that would make him chief engineer on the Erie Canal and take him to work on other canals across the nation. Little was done by the Northern Inland Lock Navigation Company, but by 1798 the Western company had completed about three miles of canals and seven locks on the Mohawk, at Little Falls and German Flats. Weston stayed two years, using his invaluable level, and Schuyler continued to direct operations.[20]

Weston's manuscript notebook has been preserved, and it gives the details of his work on the "Mohawk Lock" at Little Falls and the "Guard Lock at German Flats," including the cost and time required for digging, driving, piling, brickwork, and building lock gates. He recorded his preparation at the Mohawk Lock of mortar, mixing four and a half bushels of terras, five and three-quarter bushels of lime, and two and a half bushels of sand as he struggled with formulating a cement that would harden under water, a problem plagued every pioneer American canal and those under way in England at the time as well.[21] It was a problem Weston could not overcome.

The record of the Western Inland Lock Navigation Company in the Mohawk Valley illustrates both the failures and the hopes of the pioneer canal builders. Labor was scarce, and locks built of wood had to be rebuilt of stone, some as many as four times. Nine times the stockholders were assessed to raise funds. Merchants and boatmen protested the high tolls. The company's improvements brought significant reductions in transportation costs, but most goods still were hauled by wagon along the Mohawk, and the investors realized no profits.[22]

Still the company directors believed they were in the vanguard of the canal movement in Europe and America. Their correspondence and reports told of their hope of overcoming the rivalry of Montreal for the trade of the Great Lakes and of other routes to the interior. And as did most of the private canal companies of the period, they identified private labor with the public good.

In 1793 De Witt Clinton, who held only one share of stock but represented his uncle Governor George Clinton, charged that Schuyler was making the canal a Federalist work "to enlist the passions of a party on the side of hydraulic experiments." But the aging Schuyler remained as president until 1803, promising "to exert what remains of my ability to promoting the Interest of the W. Canal company and that of the Community, for they are mutual."[23] In 1812 the Western Inland Lock Navigation Company was taken over by the state and became part of the larger project for the Erie Canal.

These pioneer canals in Pennsylvania, along the Potomac, and in the Mohawk Valley experienced difficulties that could not be overcome, but two canals of the 1790s (in addition to the Dismal Swamp Canal) lasted more than half a century and produced substantial accomplishments. These were the Middlesex Canal, chartered in Massachusetts in 1793, and the Santee and Cooper Canal, begun a year earlier in South Carolina to bring the trade of the interior to Charleston.

The Middlesex Canal was built to link the Merrimack River, which reached north into New Hampshire, with the Charles River and the port of Boston. Meeting the Merrimack near the present location of Lowell, the canal was not quite twenty-eight miles long, but it had a great impact on the growth of Lowell as a textile center and the development of Boston. It was thirty feet wide and three and a half feet deep and would be taken as a model by the builders of the Erie Canal in New York.

The Middlesex Canal Company was chartered by an act signed by Governor John Hancock in 1793. It was the fruition of James Sullivan's idea to build a canal between the Mystic and Merrimack rivers, which would ultimately connect Boston with New Hampshire, Vermont, and Canada. Sullivan was attorney-general of Massachusetts, and he enlisted the aid of his friend Loammi Baldwin of North Woburn. Baldwin was the sheriff of Middlesex County and cultivator of the Baldwin apple. They saw the canal as a ready means of supplying Boston with lumber, building stone, pot and pearl ashes, and agricultural products. They gathered the support of a broad-based group in Medford, who petitioned the legislature in 1792 but lacked adequate financial backing. Some remained in the company after its charter in 1793, but by the end of the decade most of the shareholders were wealthy men from Boston and its environs, many of them leaders of Massachusetts government and society. The first board of directors included the Boston merchant Thomas Russell and other men of means such as John Brooks of Medford, Joseph Barrell of Charleston, and Ebenezer Storer and Andrew Craigie of Cambridge. Among the shareholders were James Winthrop and Christopher Gore and later Josiah Quincy and John Adams. Sullivan was president of the company and Baldwin was on the board of directors. They were a group much like their counterparts in New York, who, led by General Philip Schuyler, were working during these same years to improve the navigation of the Mohawk by canals.

Like the New York canal-building gentry, the Middlesex Canal shareholders faced the constant burden of assessments, $120 a share before the first six miles of canal were completed in 1797; $460 more before the entire canal was completed; a total of $610 by 1806 and $740 before the first dividends were paid in 1819. When the Middlesex Canal was opened for navigation in 1803, it had cost $500,000.[24]

Construction of this pioneer waterway was the work of the two great engineering families of New England, the Sullivans and the Baldwins. James Sullivan served the canal as promoter and president of the company. In 1808 his son John Langdon Sullivan took over the project and extended canal navigation by locks and dams up the Merrimack as far as Concord, New Hampshire. The Sullivans were examples of what Daniel H. Calhoun has called the early "proprietor-engineer," whose compensation depended upon the profits of the company. Yet James Sullivan repeatedly stressed the "public utility" of his canal and declared his desire to rise above "the engrossing pursuits of self."[25]

Loammi Baldwin directed construction before John Langdon Sullivan took over the management of the company. After initial surveys by Samuel Thompson of Woburn, which proved highly unreliable, Baldwin was named chief engineer in 1794. The canal passed by his own door in North Woburn, and he was assisted by his five sons, one of whom, Loammi, Jr., went on to become the leading civil engineer in New England. The younger Baldwin's name would be associated with canal projects from Maine to Georgia. The inadequacy of Loammi Baldwin's crude leveling instrument soon made it clear that more expert help was needed. He was directed to set out for Pennsylvania and Virginia to "visit all the canals to which you can conveniently gain access & gain all the information you can on the subject of canalling." In particular, he was to secure the services of William Weston, who was at work on the

Schuylkill and Susquehanna Canal, and, equally important, to have the use of Weston's level.[26]

Weston came for less than three weeks in 1794 and charged the company more than $2,000. (General Henry Knox had warned the company against such an occurrence, believing that foreigners were "defective in morals and steadiness.") But Weston's levels were essentially accurate, and he lent his leveling instrument until 1797. Two similar Y levels were obtained from the Troughtons in London. And after Weston departed, Baldwin wrote to him for advice when new problems arose. In 1794 he wrote on so elementary but fundamental a question as the design of a wheelbarrow: "Pray send the dimensions of one with a section figure in each direction with length, breadth and height, diameter of wheel, and whether of wood or iron."[27] Yet Baldwin drew upon his own knowledge of pumps, which he had gained studying with John Winthrop at Harvard, and he experimented on his own with hydraulic cement. In 1796 the company chartered Theodore Lyman's sloop to sail to St. Eustatius in the West Indies for forty tons of Dutch trass which Baldwin used to make his own mixture of trass, lime, and sand and which hardened underwater. Two years later he could report that his mortar was still "almost as hard as the stone itself."[28]

The Middlesex Canal was built with trial-and-error technology. Between the village of Middlesex on the Merrimack and the Charleston Mill Pond, which connected to Medford and Boston, twenty locks and eight aqueducts were constructed. They required the cutting and transporting of massive pieces of stone. The Shawheen aqueduct rested on four stone piers, was 188 feet long, and rose 35 feet above the river. But the expense of stone locks led to a change to wood in 1799, which required constant repair and rebuilding. Baldwin learned the technique of "puddling" to make the canal banks more watertight and produced a canal that leaked less than the porous banks being built at the same time on the Schuylkill Canal.

Baldwin's troubles with the canal, and with the directors, were so persistent that in March 1803 he wrote in his diary that he was "pretty much determined to quit the canal."[29] But he did not quit, and the following December, just within the ten years allotted by the company's charter of 1794, water passed from the Middlesex Canal into the Charleston Mill Pond. Only with an awareness of the repeated failures and problems overcome can we understand the slow introduction of the

Canal Era in America and reach a valid assessment of the place of internal improvement in the politics and economic development of the early American republic.

For a time, the Middlesex Canal Company operated a few boats, none too profitably, while new boating companies rose to dominate the canal trade. Freighters forty to seventy-five feet long and as wide as nine and a half feet passed through the locks and over the aqueducts bringing the savings in transportation costs always anticipated and usually realized. From Concord to Boston the down rate dropped from $8 a ton to $4 by 1838, and the rate up the canal was reduced from $13 a ton to $5. This lowered the cost of overland transportation by half, even including canal tolls.[30]

As New Yorkers discovered when they improved the navigation of the Mohawk, the competition from teamsters and wagons did not end but instead forced a continual lowering of canal tolls. But reduced costs encouraged the rise of manufacturing along the Merrimack; canal traffic increased and attracted the rival railroad, which appeared with the chartering of the Boston and Lowell Railroad in 1830. Ironically, two sons of Loammi Baldwin, James F. and George R. Baldwin, designed the railroad that ultimately killed their father's pioneering canal project. Their father's final words to his sons in 1807 had been, "Do all the good you can and cherish the Middlesex Canal."[31] In that year Loammi Baldwin, Jr., went to England to study railroad engineering. He later had a great career in American canal and railroad building.

In the first decade of the operation of the Middlesex Canal, lumber came down from New Hampshire to Boston, but there was little up-canal traffic; tolls rose from $3,300 in 1803 to $24,800 in 1814. After 1815 up-canal traffic grew and became more varied as factories multiplied on the Merrimack. By 1833, tolls reached $45,500.[32] That year the dreams of the forty-year-old company seemed finally to be realized.

In his study of New England transportation, Edward C. Kirkland has described a "golden age" for the Middlesex Canal in the 1830s. Up the canal went cotton and coal bound for the Lowell mills, along with salt, flour, and fish. Down the canal came cotton cloth, iron, copper, and bricks. There were passenger packets as well. Edward Everett Hale later wrote about his 1826 ride as a boy on the packet the *General Sullivan* up the canal to Chelmsford on the Merrimack. "To sit on the deck of a boat and see the country slide by you, without the slightest jar, without a cinder or a speck of dust," was, he remembered, "one of the exquisite luxuries."[33]

Canal boats dominated the river traffic of the Merrimack, poled upstream or aided by sails, often carrying the bricks from which the new factories were built. It was such boats that Henry Thoreau witnessed on the Merrimack in 1839, when he saw "canal-boats, at intervals of a quarter of a mile, standing up to Hooksett with a light breeze, and one by one disappeared round a point above. . . . It was a grand motion, so slow and stately." But Thoreau thought the Middlesex Canal had "even an antique look beside the more modern-railroads," which were just then appearing alongside the canal.[34]

But with the completion of the Boston and Lowell Railroad in 1835, the Nashua and Lowell Railroad in 1838, and the addition of the Concord and Nashua Railroad after 1842, the rails took trade from the canal over the entire route from Boston to Concord, New Hampshire. In the last decade of the canal's operation, from 1843 to 1853, declining traffic consisted mainly of lumber products and toll receipts barely exceeded expenses. There were no profits for the company's investors, who received dividends of only $559 per share between 1819 and 1853.[35]

The company's achievement had been to provide, with remarkable persistence, support for a venture that promised a private gain which could not be realized and also bringing lumber from New Hampshire to Boston and furnishing the essentials for the great Merrimack mills. Perhaps most important, the Middlesex Canal demonstrated a new transportation technology to the nation.

While Loammi Baldwin struggled to construct his canal in Massachusetts and Philip Schuyler toiled to build locks along the Mohawk, in the deep South the Swedish engineer John Christian Senf worked to build another pioneer waterway, the Santee and Cooper Canal in South Carolina. That canal resulted from the effort to connect the Santee River, which drained the upland agricultural region of South Carolina, to the Cooper River, which reached northwest from Charleston.

The Santee entered the Atlantic fifty miles above Charleston. Near its mouth it followed a twisting course through swampy lowlands, a hindrance that could be overcome by a canal to the Cooper River, which flowed directly to Charleston Harbor. South Carolina planters knew of the success of English canals and the pioneer canals proposed in the

North and had long used small canals and ditches to drain and control the water in their rice fields, which sustained the culture of "Carolina Golden Rice."[36] It was apparent that a short canal between the Santee and Cooper rivers would give upcountry planters greater access to the Charleston market and bring cheaper foodstuffs from the uplands to that city.

The collapse of indigo production after the removal of the British bounty, which had been given before the Revolution, spurred Charleston investors to improve transportation to the interior. A charter for the Santee Canal Company was secured in 1786. Again it was a planter-commercial gentry who led in canal building. The company included men such as Governor Francis Moultrie, John Rutledge, William Henry Drayton, Thomas Pinckney, General Thomas Sumter, General Francis Marion, and Henry Laurens.[37] Construction on the Santee Canal began in 1792, and the twenty-two-mile waterway with its twelve locks and eight aqueducts was completed in 1800. The canal left the Cooper River, which was navigable for thirty miles above Charleston, and ran almost due north to the Santee River three miles west of Pineville.

If William Weston and Benjamin Latrobe brought their near indispensable engineering knowledge to northern canals in the 1790s, the direction of the Santee and Cooper Canal by John Christian Senf was a mixed blessing. Senf had been captured along with General John Burgoyne's Hessian troops at Saratoga and was sent to South Carolina by Henry Laurens. General Horatio Gates said in 1780 that he was "the best Draughtsman I Know, and an excellent engineer," and Jefferson, governor of Virginia during the Revolution, wanted Senf as state engineer for Virginia. Senf assisted in the fortifications at Hoods, Virginia, and Jefferson recommended him as "a Gentleman eminent for his Skill as an Engineer, his Zeal and activity."[38]

Senf, a vain and forceful personality, disregarded the earlier suggestions of Henry Mouzon that the canal follow a natural Santee tributary stream for its water supply. Instead, he designed the canal to cross the twelve-mile-long dry Santee ridge, which rose thirty-four feet above the Santee and sixty-nine feet above the Cooper River. Senf's canal was thirty-five feet wide and four feet deep, had locks sixty feet long and ten feet wide built largely of brick, and used reservoirs built to fill the dry portion on the summit level.

That so short a work required eight years for construction reveals again all the difficulties of the new canal technology in America.

Moreover, the canal crossed some thirty-five plantations to reach the Cooper River, and many of the owners feared it would take water from their rice fields and that passing canal traffic would have a bad influence on slaves kept heretofore in isolated labor.

Slaves furnished most of the labor to build the canal, their numbers reflecting the vicissitudes of the South Carolina economy. The post-revolutionary war depression made surplus slave labor available, but cotton expansion in the 1790s following the introduction of the cotton gin raised the cost of hiring slaves from their owners. Six hundred laborers, mostly slaves, were engaged in 1796, and the company purchased some sixty slaves. They were particularly essential as the major source of bricklayers. Some of the shareholders met the repeated assessments by hiring out their own slaves to the company. With help from a legislative grant for a lottery in 1796, shareholders were pushed to the limit to meet the $750,000 total cost of the canal.[39]

In 1800 a cargo of salt was carried through the canal and up the Congaree River to the town of Granby at the foot of the rapids below Columbia, and the next year the *Charleston Times* reported a boat sent by a planter on the Broad River, ninety miles above Grandy, "having safely passed over all the falls, and shoals," by canal to Charleston.[40] The rapids on the Congaree at Granby were soon passed by the Saluda Canal.

Tolls on the Santee Canal were first determined by the width of a boat, from $10 for a boat six-and-a-half-feet wide to $25 for an eight-and-a-half-foot one. But by 1821 tolls were set at $15 a boat, most of which carried twenty-five tons of cargo, or 120 bales of cotton, at about half the cost of land carriage. Almost $13,000 was collected by the company in tolls in 1807, but no dividends had yet been paid.[41] Unlike most canals of the time, steam towboats were used on the Santee. In 1822 the Charleston and Columbia Steam Boat Company operated a fleet of towboats through the canal and on the upland rivers. In that year the *Carolina* towed boats carrying 900 bales of cotton from Granby through to Charleston in only four and a half days.[42]

The state of South Carolina built or aided a series of short canals to improve the navigation of its upland river system. In 1787 the Company for Opening the Navigation of the Catawba and Wateree Rivers was chartered, and others were organized for improving the Broad River and connecting the Edisto and Ashley rivers, but each of these failed. By 1826 the state had built a series of short canals which allowed boats to come down the Catawba River to the Santee Canal and Charleston.

Canals from the Santee to the Saluda and to the Broad River opened navigation to Columbia.[43]

Although it was begun to bring food from the upland regions, the Santee Canal increasingly carried cotton, following the widespread application of the cotton gin in the 1790s. As many as seventy thousand bales of cotton came down the canal each year, although because of its light weight and high value, cotton could be carried profitably by team and wagon. But when the railroad from Charleston to Columbia opened in 1842 (then the longest in the nation), canal traffic rapidly declined. Cars carrying cotton to Charleston were pulled by the first locomotive built in America, the Best Friend, made at West Point, New York. Like the Middlesex Canal, after 1850 tolls barely exceeded expenses, and in 1858 the Santee Canal was closed.[44]

Almost as much as the pioneer canals in the North, the Santee Canal heralded the Canal Age in America, even though it was not as long as the Middlesex Canal, had fewer structures, and had less economic impact on its region. The Santee Canal was part of a visionary plan by Robert Mills for a water route from Charleston to the Mississippi River and beyond. Mills was born in 1781 in Charleston, was educated at the College of Charleston, moved to Washington, Philadelphia, and Baltimore, and lived for two years at Monticello with Jefferson, with whom he shared an interest in canals and the West. Returning to South Carolina in 1820, Mills joined Joel Poinsett on the Board of Public Works and published a plan for a canal from Charleston to Columbia which would ultimately be part of a great western water route to the interior. His famous *Atlas* of 1825 detailed every canal in South Carolina, and his *Treatise on Inland Navigation*, published in Baltimore in 1820, proposed a "vast internal navigation" by canals and surveyed the waterways of the nation that might be linked to Baltimore on the Atlantic coast.[45]

The private canal companies worked against almost insuperable obstacles with scanty funds to develop transportation routes that were often national in scope, and their difficulties clearly demonstrated the need for national assistance. To be sure, most of the early privately sponsored canals attracted limited state aid.[46] Even if national support was not provided, there was little overt opposition to federally supported internal improvements on constitutional grounds. Indeed, the opposite was true. Canals and other internal improvements were widely regarded as essential to the preservation of republicanism.

The stimulus of a "republican technology" in which the "ideology of republicanism helped to provide a receptive climate for technological adaptation and innovation" has been identified by John F. Kasson.[47] It is illustrated repeatedly on the canals of the 1790s. A petition from fifty-six men in Woburn in 1795 to the directors of the Middlesex Canal expressed the desire that the canal be located on their side of the village. But they also described the canal as "calculated to promote the public interest and the welfare of every individual," and they attested to their belief that "the public Good will be your Polar-star in determining the question."[48] Concern for the public good dotted the writings of canal builders in the new nation.

Republican values underlay the emerging use of the power of eminent domain. This power was given to the proprietors of the Middlesex Canal in the incorporating act of 1794 and to the Schuylkill Navigation Company in 1792. A remonstrance of farmers in the Lebanon Valley in 1793, however, makes it clear that this power was not yet clearly established. Appealing to "the inalienable rights of the people" and "to the enlightened understanding of the representatives of a free people," the protesting farmers prayed " to have those obnoxious laws repealed by the authority of which companies have been incorporated to dig canals and make turnpike roads, to the injury of public and private property."[49] These farmers directed their attack against "a few wealthy and powerful citizens . . . authorized to dig Canals," but already in the 1790s there were indications that when larger canal ventures were attempted they would be under state ownership. In 1791 the memorial of the Society for Promoting the Improvement of Roads and Inland Navigation in Philadelphia proposed a grand water route to the Ohio and Lake Erie with the clear conviction that it should be done by the state.[50]

Federal aid to internal improvements has long been assumed to have been a cornerstone of Federalist economic doctrine. In Washington's administrations in the 1790s, Alexander Hamilton gave cautious approval to the proposals of Treasury Department Assistant Secretary Tench Coxe for federal construction of roads and canals. But Hamilton's first concern was for his fiscal policy and American trade with Great Britain, and he feared local jealousies in the adoption of any comprehensive plan for internal improvements. In his Report on Manufactures in 1791 he noted "the symptoms of attention to the Improvement of inland navigation" recently displayed, which would "stimulate the exertion of the Government and Citizens of every state." But he was surprisingly hesitant when he wrote that "it were to be wished that there was no doubt of the power of the national government to lend its direct aid, on a comprehensive plan."[51]

Tench Coxe was the most ardent advocate of federal support for internal improvements in the early 1790s. Like Washington, Coxe speculated in western lands, and he saw roads and canals as essential for commerce to the "Western Country." He bought land in New York, western Pennsylvania, and North Carolina. His lands in the Wyoming Valley in Pennsylvania would later be developed for coal and iron, enriching his heirs. Far beyond his personal interest, Coxe wrote "tens of thousands of words," as Jacob Cooke has examined them, to popularize the concept of an interdependent agricultural, manufacturing, and commercial economy in the national consciousness.[52] His zeal resembled that of Henry Clay in his "American System."

Though Coxe put greatest emphasis on manufacturing, internal improvements on a large scale were a central element in his vision of what the United States might become and the prospects of discord or disunion that he feared. He was interested in the Dismal Swamp Canal, proposed other canals, and was a member of the Pennsylvania Society for the Improvement of Roads and Inland Navigation. Writing to Jefferson in 1801, Coxe described his labors as sacrifices "in the *great and certain struggle* to which Republicanism is yet *destined* in America."[53]

Although the companies that built the pioneer canals were nonpartisan in sponsorship, it was Federalists such as Philip Schuyler in New York, Henry Knox in Massachusetts, and George Washington who were often in the lead. Robert Morris was president of the Susquehanna and Schuylkill Navigation and the Conewago Canal Company. Far down the Ohio River, Federalists from New England dominated the Ohio Company's settlement at Marietta and included plans for canals in their efforts to link their western outpost to the East.[54]

But the Jeffersonians James Madison, Albert Gallatin, and Jefferson were no less active in their advocacy of internal improvements. In particular, they sought to preserve republican institutions by agrarian expansion to the West, and in the end they gave the greatest stimulus to internal improvements in the first decade of the new nation. Drew McCoy has described "Madison's endeavor to preserve and expand across space the youthful, predominantly agricultural character of American society."[55] Madison had worked for the Potomac Company in the Virginia legislature, and he wrote in *Federalist* No. 42 that "nothing which tends to facilitate the intercourse between the States can be deemed unworthy of the public care." Albert Gallatin advanced a political economy in opposition to Hamilton in the 1790s, but he sought equal access to a national market system in his proposals in 1802 for internal improvements submitted to William B. Giles in the House of Representatives. Gallatin wished to apply 10 percent of the proceeds from land sales (similar to a plan of Tench Coxe, who was now a Jeffersonian) to build turnpikes connecting with navigable waterways, and his annual Treasury reports of 1806 and 1807 began to stress the advantages of internal improvements to strengthen the bonds of union.[56]

Far from opposing internal improvements out of fear that they would require national assistance and strain his belief in a strict construction of the Constitution, Jefferson was no less supportive of internal improvements. He was proud of having been the first to propose the improvement for navigation of the Rivanna River, which ran by his door in Virginia and emptied into the James River. He had supported the Potomac Canal and urged the Potomac Company to consider building a canal far to the northwest between the Beaver River and the Cuyahoga River, which flowed north to Lake Erie. He believed that the Dismal Swamp Canal should be built at public expense. In 1787, riding in a glassed-in carriage mounted on a canal barge, he had spent a week traversing the whole of the Languedoc Canal in France and had sent notes to Washington describing this longest canal in Europe. He was fascinated by canal locks and designed a sluice gate.[57]

Jefferson's emphasis on economy and minimal government renders any generalization difficult, but he moved increasingly toward a nationalistic, mercantilistic use of federal power for internal improvements. John L. Larson has argued that what Jefferson feared most was logrolling and the pork barrel, corruption in government rather than giving government the power to build internal improvements.[58] In proposing a comprehensive program of internal improvements, the Jeffersonians would go far beyond the economic plans of Washington and Hamilton.

In his second inaugural address in 1805, Jefferson noted the fruits of his governmental economy, the imminent retirement of the national debt, and the continuing revenue from tariffs. He then raised the possibility that surplus revenues, by "a corresponding amendment of the Constitution [might] be applied *in time of peace* to rivers, canals, roads, arts, manufactures, education, and other great objects within each State." Madison, who was especially concerned about preserving republicanism by augmenting the commerce of the agricultural West, asked, "What is the amendment alluded to?" The following year, in 1806, Jefferson's annual message repeated his plea for a constitutional amendment to use the impost for "the great purposes of the public education, roads, rivers, canals, and such other objects of public improvement."[59]

Canal building was also on the minds of Jefferson's associates and supporters whose ideas fed in varying degrees the movement that produced Albert Gallatin's *Report of the Secretary of the Treasury on the Subject of Roads and Canals* in 1808. Jefferson read with approbation Joel Barlow's poem *The Vision of Columbus*, in which Barlow wrote, "Canals, long-winding, ope a watery flight / And distant streams and seas and lakes unite," while "New paths, unfolding, lead their watery pride / and towns and empires rise along their side." Barlow joined Jefferson in Washington in 1806 and the next year published the *Columbiad*, which again linked roads and waterways with republican virtue. In New York, the Jeffersonian Robert Livingston joined in a steamboat monopoly with Robert Fulton, whose 1796 pamphlet set forth the particular virtue of canal navigation. Closely associated with them was Turner Carmac, who published a pamphlet on "the great utility of an extensive plan of Inland Navigation in America," and the young civil engineer Benjamin Latrobe.[60]

Lee W. Formwalt has found Benjamin Latrobe to be the most influential in shaping Gallatin's and Jefferson's planning for internal improvements. Latrobe was serving as architect of the national Capitol, and like Fulton and Barlow he was a close friend of Jefferson. He had been born in England and educated in Germany and had admired the German and Austrian economic improvement programs of the cameralists. He enhanced his knowledge of civil engineering in England and in 1801 and 1802 brought his skills to the Susquehanna Canal, which directed the Susquehanna River trade toward an outlet in Baltimore. Latrobe was then chosen by the Philadelphia backers of the Chesapeake and Delaware Canal in 1804 to begin work on the feeder for the canal summit, which effort he pursued until the financial collapse of the canal company in 1805. It was the campaign for federal aid to the Chesapeake and Delaware Canal that led to Gallatin's famous report of 1808.[61]

Philadelphia merchants had long wanted a canal to cross the Delmarva Peninsula between Chesapeake Bay and Delaware Bay. In Congress, James Bayard of Delaware and Joshua Gilpin of Pennsylvania introduced a plan in 1806 to aid the Chesapeake and Delaware Canal, but it failed. Then, in his sixth annual message in 1806, Jefferson suggested that public revenues be used for "roads, rivers, canals, and such other objects of public improvement," and Henry Clay of Kentucky offered the prospect of combining aid to the Louisville and Portland Canal Company with the project for the Chesapeake and Delaware Canal. Clay spoke in favor of a land grant for a canal around the falls of the Ohio, and Bayard called for a land grant to revive the Chesapeake project. Thomas Worthington of Ohio proposed land grants to both improvements.

But John Quincy Adams spoke strongly against both projects, in sharp contrast to his later support of internal improvements, accusing the senators of "combining to divide public lands, and public treasuries." Adams called for the secretary of the treasury to report on a plan for roads and canals which would merit the aid of the national government, and on Worthington's motion Adams's resolution passed the Senate.[62] The result was Gallatin's *Report on Roads and Canals* of 1808.

Gallatin's report was a great blueprint for a national system of internal improvements, which Carter Goodrich has called "one of the great planning documents in American history." It has been portrayed as Gallatin's effort to rise above the clamor for competing state projects, for which influence in Congress might be traded and drain the public treasury, and to substitute the logic and authority of a cohesive national plan. Yet others have described Gallatin's plan as an instrument for Jeffersonian development of a national market system necessary for preservation of republicanism and the Union.[63]

Gallatin's report projected four great canals along the eastern coast to cut across the Atlantic capes and necks and improve navigation inland: the Cape Cod canal, the Delaware and Raritan Canal in New Jersey, the Chesapeake and Delaware Canal, and the Dismal Swamp Canal, already begun between the Chesapeake and Albermarle Sound.

Gallatin recommended the building of canals across the Appalachians to improve "four great Atlantic rivers," or rather, river systems—the Allegheny-Juniata and Susquehanna, the Monongahela and Potomac, the Kanawha and James, and the Tennessee-Savannah and Santee rivers—with roads across the mountains to connect them with western rivers. But only in New York could a canal break through the Appalachian chain and connect the Hudson River with the Great Lakes, and Gallatin projected part of the future Erie Canal to Lake Ontario with an additional canal around the falls of the Niagara River. Gallatin endorsed the canal project around the falls of the Ohio and other smaller waterways as well. The canal projects already begun along these routes, the new ones proposed, and the roads to be added were part of a great national program, which Gallatin estimated might be completed in ten years at a cost of $20 million. Such estimates were overly optimistic, but his report included detailed communications from knowledgeable local people and an appendix containing authoritative information on most of the turnpikes and canals in America.[64]

Most of the improvements proposed in Gallatin's report were finally constructed, though some by railroad rather than by road or canal. The time required would stretch to fifty years and more. Politically, Gallatin was as far from the mark as he had been in his estimates of time and cost. Only three of his proposed waterways were assisted by the national government; a few were completed by private funds; some were built by the states alone; and most received mixed support from state governments, local governments, and private companies.

Although Jefferson preferred to wait for a constitutional amendment to implement Gallatin's plan, Congress almost immediately received a host of petitions from states seeking national aid. In 1810, Senator John Pope of Kentucky sponsored a comprehensive bill written by Benjamin Latrobe, and in the House, Peter B. Porter of western New York submitted a similar measure. Together these bills embodied the Gallatin proposals but added canals to link the Allegheny River and Lake Erie, to join the Susquehanna and Delaware rivers, and to link the Roanoke and Appomattox rivers.[65]

Porter spoke for the state of New York, which had already gone far in preparing its plans for canals to Lakes Erie and Ontario, as he argued the necessity for "a great navigable canal from the Atlantic to the Western states." Canal commissioners in New York had already drawn up reports projecting the Erie Canal, and they sent Gouverneur Morris and De Witt Clinton to Washington in 1811 to lobby for national assistance. Their appeal was sweetened by the addition of other projects as planned by Gallatin and Latrobe, but Congress refused its support. Troubled commercial relations abroad and the War of 1812 killed any immediate implementation of the Gallatin-Latrobe plans, but there was also opposition from a block of New England Federalists, southern conservative Republicans, and an anti-Gallatin faction in the Senate.

The end of the War of 1812, however, was followed by the beginning of a great era of American canal building. As Joseph H. Harrison has written, "Demands for internal improvements mounted, gaining in intensity from the very circumstances that postponed their fulfillment."[66]

Especially in the mid-Atlantic states and along the coast the travails of war had dramatized the need for better roads or canals. The great burst of canal construction would begin with the Erie Canal in 1817.

Yet the promise of federal aid that seemed imminent in Gallatin's report and in the reports of Senator Pope and Representative Porter never reached fulfillment. Federal aid to canals was caught in the deep cross-currents of national opinion, presidential action, and congressional policy. Paradoxically, the emergent postwar nationalism yielded to an era of state power, and the states moved forward with canal building.

Madison's earlier nationalisitc support for internal improvements appeared to find new strength in his annual message of 1815 in which he stressed "the great importance of establishing throughout our country the roads and canals which can best be executed under the national authority." His tone was ardent: "No objects within the circle of political economy so richly repay the expense bestowed on them; there are none the utility of which is more universally ascertained and acknowledged. . . . Nor is there any country which presents a field where nature invites more the art of man to complete her own work for his accommodation and benefit." Although states might build "navigable canals," the "General Government is the more urged to similar undertakings . . . by the prospect of thus systematically completing so inestimable a work." But Madison allowed a shadow to cloud his bright picture of assistance to such undertakings by qualifying his reference to those "requiring a national jurisdiction and national means"; he touched on the possibility of a "defect of constitutional authority," which he dismissed with "a happy reflection" on the "mode which the Constitution itself has providentially pointed out." Like Gallatin and Jefferson before him, he saw a recourse to a constitutional amendment.[67]

How much was glow and how much was substance in the postwar nationalism which promised so much for a national system of roads and canals? Although a new literature celebrates concern for the preservation of republicanism in the new nation, and Roger H. Brown has portrayed the War of 1812 as a struggle to save a "republic in peril," Robert Wiebe has described "an impotent national government" barely saved from collapse. For all the rhetoric of nationalism, wrote Wiebe, "the national government shrank into inconsequence."[68]

If such atrophy was indeed occurring, it did not dim the work of congressional committees preparing for national aid for internal improvements, especially canals. Madison repeated his nationalistic exhortations to the Congress that met in December 1816, reminding the legislators of "the expediency of exercising their existing powers, and where necessary, resorting to the prescribed means of enlarging them, in order to effectuate a comprehensive system of roads and canals." Prospects for the nationalistic canal proposals of Gallatin, Latrobe, Pope, and Porter appeared to be alive and well.[69]

What emerged from Congress was John C. Calhoun's Bonus Bill, which would use the bonus and dividends paid to the national government by the Second Bank of the United States as a fund for internal improvements to be allocated to the states. To be sure, the protracted debate included the issue of constitutional authority, but Calhoun replied to his opponents: "We are great and rapidly—[I] was about to say fearfully—growing. This . . . is our pride and danger, our weakness and our strength. . . . Let us then . . . bind the Republic together with a perfect system of roads and canals. Let us conquer space." For the moment at least, Calhoun's phrase captures the nationalist spirit of the age. In Congress the opposition was defeated; the Bonus Bill squeaked through the House by a margin of two and succeeded in the Senate as well.[70]

But Calhoun and the supporters of the Bonus Bill in the Fourteenth Congress were soon stunned by the surprising veto of James Madison, who had remained silent during the congressional debate. Henry Clay could not believe Madison had abandoned his earlier nationalism. When the issue was debated again in 1818, Clay appealed "to the members of the last Congress" to attest that "no circumstance, not even an earthquake that should have swallowed up one half of this city, could have excited more surprise than when it was first communicated to this House, that Mr. Madison had rejected his own bill—I say his own bill—for his Message at the opening of the session meant nothing if it did not recommend such an exercise of power as was contained in that bill." The two messages, said Clay, reading to the House Madison's message at the beginning of the session and then his veto message, were "perfectly irreconcilable."[71] Indeed, Madison later wrote to his successor in the presidency, James Monroe, "The expedience of vesting in Congress a power as to roads and canals, I have never doubted."[72]

But in his veto message of 1817, Madison explained his rejection on constitutional grounds, arguing that the power of the national legislature to provide for roads and canals "is not expressly given by the Constitution." What had been only a shadow on his earlier nationalism regarding internal improvements had become a prohibitive reservation. Madison announced that "the permanent success of the Constitution depends on a definition of powers between the General and State governments, and that no adequate landmarks would be left by the constructive extensions of the powers of Congress as proposed in the Bill."[73] Therefore he could not sign it.

The real reason for Madison's veto may have arisen from the strong opposition of Virginia leaders, who feared the rivalry of New York, which stood ready with its plans for an Erie Canal. Concessions made under duress to secure passage of the Bonus Bill may have made the measure a bad bill. For example, congressional appropriations preceded the definition of which roads and canals were to be constructed. A requirement of state consent limited federal authority. Development of a national system was threatened by the requirement for proportional spending, and the proponents of the Bonus Bill knew that any attempt at further definition of details would only produce a self-defeating conflict over rival improvements.

Although most historians have found in Madison's constitutional objections the resurgence of his strict constructionist Jeffersonian convictions, John L. Larson has found the key to Madison's veto in his republican fear of corruption and logrolling. In this interpretation, Madison feared that states would plunder the treasury, factions would arise, and the balance of power under the Constitution would flow excessively toward the Congress.[74]

The long debates over the Bonus Bill in 1817 and those in 1818 over the power of the national government to construct internal improvements gave ample evidence of state rivalry and regional conflict. The localism always indigenous to internal improvements surfaced as state delegations in Congress divided for and against, even among the congressmen from New York and Virginia. Opponents of national assistance charged that New York's Erie Canal would be the largest beneficiary of such aid, to which Representative Thomas R. Gold from Oneida County, New York, responded, "It has been . . . the singular lot of great national enterprises, and especially of canals, to be resisted by prejudice, by narrow calculations and short-sighted views." Yet, he concluded, "It is, sir, by improved roads and canals that commercial intercourse . . . is promoted," and only by such improvements could the great sections of the nation "be cemented and preserved in a lasting bond of union."[75] As the Bonus Bill was being debated in 1817, Thomas Wilson of Erie, Pennsylvania, noted the projects in other states ready to go forward—the Chesapeake and Delaware Canal, the Dismal Swamp Canal, projects in Pennsylvania and New Jersey, and the canal around the falls of the Ohio at Louisville—which awaited national support.

The issue of constitutional authority was at the forefront of these debates, whether as a genuine question of philosophy and interpretation or as a cover for political and economic interests. So Thomas B. Robertson of Louisiana wanted internal improvements left to the states, improvements that were "peculiarly their province, unless indeed we are to have what is anxiously wished by many—one grand, magnificent, consolidated empire."[76] Calhoun's astute political management in 1817 and Clay's winning speeches in the debates of 1818 carried the argument for nationalistic aid to internal improvements, and together they were the bulwark against the opposition. For Calhoun and Clay the Constitution gave a flexible framework under which to provide for the growth of a living, expanding nation.

As the congressional internal improvement debates of 1818 moved largely to the question of constitutional authority, the persistent themes of the preservation of the Union and republican government were heard again and again. Henry Clay feared that the settlers of the Mississippi Valley would lose their "moral" attachment to the Union unless the "General Government" penetrated "the intervening mountains by roads" and connected their navigable waters "by such links as, for example, the great canal of New York."[77]

John P. Cushman of Troy, New York, warned that the trade of the Great Lakes and New England would be siphoned to Montreal, and against this possibility he asserted, "canals are the only permanent remedy." New York's Erie Canal, he said, was "one of the proudest monuments of national glory" and he pointed to the "national interests it is destined to subserve."[78]

Similar nationalistic arguments for internal improvements came from Archibald Austin and Charles Fenton Mercer of Virginia and Francis Jones of Tennessee. The fires of postwar nationalism were by no means extinguished. When the prolonged debates over the power of Congress to finance internal improvements wore down to a close in the Fifteenth Congress, Representative William Lowndes of South Carolina told his colleagues that there could be no doubt that a large majority of the House believed in "the power of Congress to appropriate money for the constructing of roads and canals."[79]

The principle of national aid to internal improvements had been sustained, but the most that could be achieved were resolutions to request reports from the secretaries of war and the treasury for further information and planning. Direct assistance to specific projects was yet to come. New Yorkers, disappointed in their hopes for national assistance, were especially bitter.

As De Witt Clinton observed, "After swallowing the National Bank and the Cumberland Road it was not supposed that Mr. Madison would strain at Canals, but so it is."[80] In 1817, New York quickly decided to go on with its Erie Canal alone. Most of the other states that built canals would do the same, with no or very limited national assistance. A Delaware editor watching the failure of the Bonus Bill and anxious for federal aid to the Chesapeake and Delaware Canal wrote that responsibility for internal improvements was thrown "back upon the people, and the respective states. If *Uncle Sam* cannot help us, we must help ourselves."[81]

If their rhetoric can be believed, the states reflected the same themes when they won popular support for their projects, asserting their belief in national goals and republican values. Congressional defeat notwithstanding, nationalism and republicanism persisted as vibrant forces in the states, even as they financed their own projects. Some canals such as the Chesapeake and Delaware and the Chesapeake and Ohio would yet win congressional assistance. But for most canals, the issue of national internal improvements continued only in the realm of congressional debate while the states sought to use their canals to preserve the values of republicanism. The Bonus Bill veto of 1817 was, as Robert Wiebe has written, part of a transition to an "era of state power."[82] The great burst of canal building that began in New York with the Erie Canal and was soon emulated by other eastern states in their attempts to surmount the Appalachian barrier to the West would be largely the work of the states.

The origins of most of these state canals lay in the pioneer waterways of the 1790s, some of which operated well into the nineteenth century. In New York the history of the Erie Canal began with the pioneer Western and Northern Inland Lock Navigation companies, and the dimensions of the great canal being built across the length of New York were copied from the pioneer Middlesex Canal still operating in New England. Going it alone, New York made the Erie Canal the outstanding success of the Canal Era.

2

Great Lakes to Atlantic: Canals of New York and New England

The Erie Canal in New York was the first to demonstrate on a grand scale that canals would work in America. In its first years it was popularly known as the "Grand Canal." Begun in 1817, finished in 1825, the Erie Canal crossed overland 363 miles from the Hudson to Lake Erie, winding its way up the Mohawk Valley and breaking through the Appalachian chain. Just as it took the pioneer Middlesex Canal in Massachusetts for its model, the Erie Canal became the model for most of the subsequent canals in America.

Canal technology was exported to other states. New York engineers became itinerant canal builders, adding an organizational loyalty to the earlier, more individualistic profession of American civil engineering.[1] The success of the Erie Canal provoked almost immediate imitation, even in places where such favorable conditions for canal building did not exist or where there was nothing to match the Mohawk gap in the Appalachian chain. To western states the Erie Canal offered a water connection to the Atlantic. To New York City it promised a competitive edge that threatened every other Atlantic seaport. It shaped the concept of New York as the Empire State. As David M. Ellis has written, "Perhaps the outstanding achievement, simultaneously the symbol and reality of New York's stature, was the Erie Canal."[2] Yet it was simultaneously a symbol of nationalism and a manifestation of republicanism.

The tiny pioneer canals of the Western Inland Lock Navigation Company furnished the kernel from which the Erie Canal grew. The company tried to improve the navigation of the Mohawk River and open a water route along the Oswego River to Lake Ontario. The goals of the company were those of the later Erie Canal: to develop western New York, to draw the trade of the West to the Hudson, and to divert that trade away from the outlet of Lake Ontario, which passed down the St. Lawrence River to Montreal.

The company's shareholders expressed a theme that was repeated over and over again in the Canal Era—the congruence of public good and private gain. Yet the failure of the company's efforts demonstrated that such a work was beyond the reach of private enterprise. Among the directors and shareholders of the company were men who would continue their work when it had become the project of the state of New York. De Witt Clinton, Elkanah Watson, Robert Troup, Thomas Eddy, Jeremiah and Stephen Van Rensselaer, Gouverneur Morris, Benjamin Wright, Gideon Granger, Jonas Platt, and Simeon De Witt all would continue their involvement with a western canal after it was no longer a canal to Lake Ontario but the staggering challenge of an overland canal to Lake Erie.

Gouverneur Morris, Elkanah Watson, and others talked or wrote in their letters about a future canal to Lake Erie, which they described in very general terms, probably thinking of a canal to Oswego and Lake Ontario with another around Niagara Falls. But it was Jesse Hawley, a western New York merchant confined as a debtor in Canandaigua, who published a series of essays in 1807 and 1808 proposing an overland "Genesee Canal." Hawley had fled to Pittsburgh in 1806 to escape debts incurred with a business partner and there published an essay in the *Pittsburgh Commonwealth* promoting such a canal. Returning to New York, he was given a twenty-month confinement "within the gaol limits" of Genesee County. From his "Debtor's Prison" he wrote the remainder of the essays that made him the first publicist for the Erie Canal.[3]

Hawley wrote fourteen essays in the *Genesee Messenger* giving a plan for "a canal from the foot of Lake Erie into the Mohawk." Until he died in 1841, Hawley would claim that the idea for such a canal was no one's but his own, and the finished canal followed a route close to the one he had marked out. Like Morris before him, however, he envisioned a canal using an inclined plane, "pitched and gauged to any dimension required," which would fall from the height of the Niagara escarpment rather than the system of lockage from level to level that prevailed in the Canal Era.[4] Along the Mohawk Hawley would rely only on river improvements.

Like the idea of Manifest Destiny which found expression in the next generation, Hawley attributed the route for his overland canal to the hand of God. In creating the "reservoir of Lake Erie" above the Niagara escarpment, "the Author of Nature . . . had in prospect a large and valuable canal . . . to be completed at some period in the history of

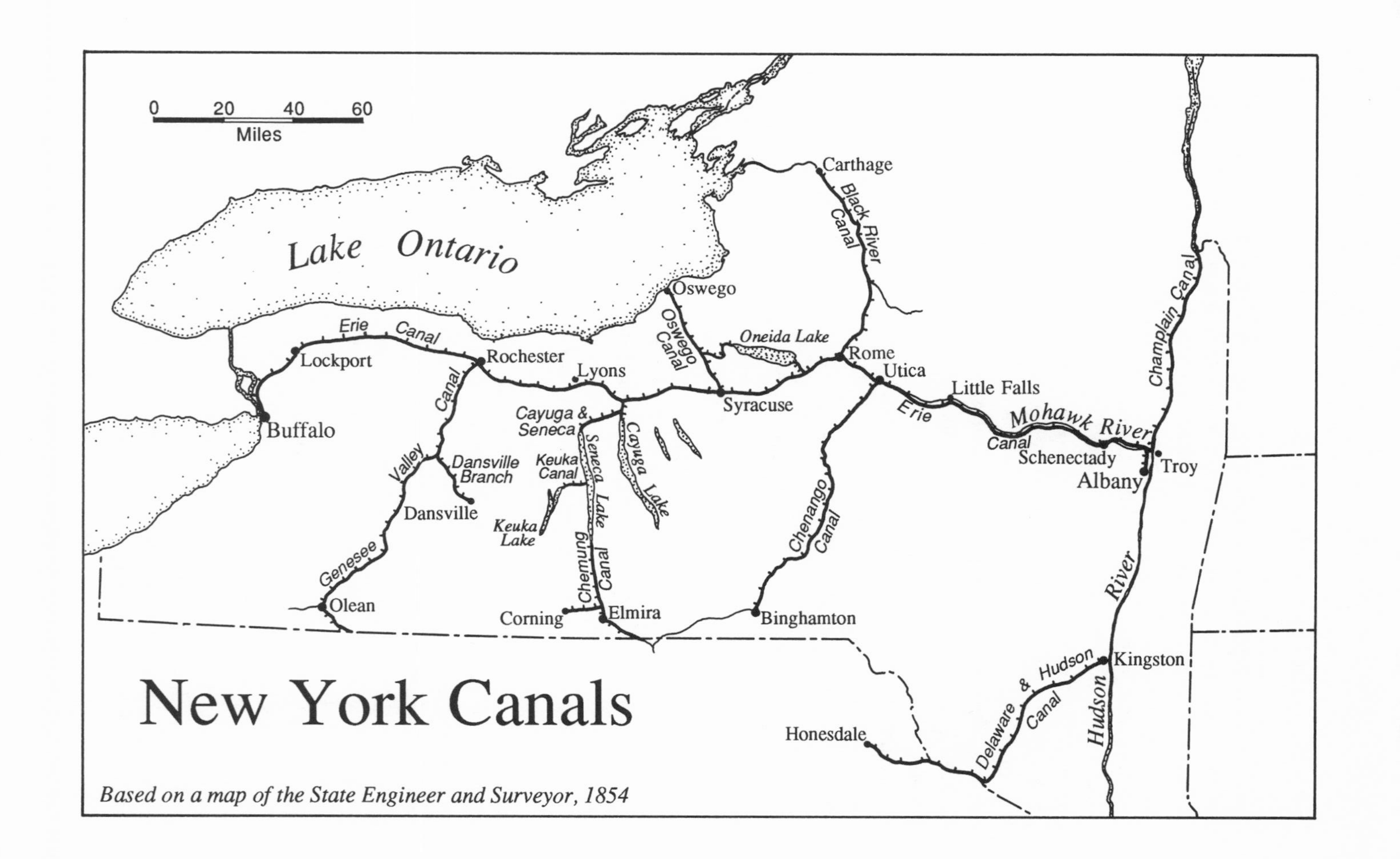
0
20
40
60
Miles
Lake Ontario
Carthage
Black River Canal
Oswego
Oswego Canal
Oneida Lake
Rome
Utica
Little Falls
Mohawk River
Champlain Canal
Erie Canal
Lockport
Rochester
Lyons
Syracuse
Buffalo
Cayuga & Seneca
Seneca Lake
Cayuga Lake
Keuka Canal
Keuka Lake
Dansville Branch
Dansville
Genesee Valley Canal
Olean
Chemung Canal
Corning
Elmira
Chenango Canal
Binghamton
Schenectady
Albany
Troy
Hudson River
Kingston
Delaware & Hudson Canal
Honesdale
New York Canals
Based on a map of the State Engineer and Surveyor, 1854

man."[5] Hawley thought he might be writing "for a subsequent age," but in five years he was applying to De Witt Clinton for a position as an engineer on the Erie Canal. And in 1825 he made the celebratory speech in Buffalo that sent the first boat through the completed canal on the way to elaborate ceremonies off Sandy Hook.

The canals of the Canal Era would follow the plans of the Gallatin report, which was published in the same year as Hawley's concluding essays. But many were also charted by Hawley, and he emphasized those that would link to his Genesee Canal. Hawley's proposals exuded the same spirit of republicanism associated with the pioneer canals of the 1790s. Only a "patriotic government, with a productive revenue," wrote Hawley, could be relied upon to build his overland canal, which would bind the Union together and preserve American republicanism. He thought the undertaking beyond the capacity of individual capital, and he warned that "the government which creates charter-parties" sacrificed its jurisdiction and "erects so many demi-sovereigns within itself."[6]

The New York legislature responded to this and other appeals from western New York with a $600 appropriation in 1808 for a survey, to be divided between the better-known Ontario route and the long interior route to Lake Erie. Not surprisingly, Simeon De Witt, the surveyor general, sought the interest of the Holland Land Company and its agent Joseph Ellicott at Batavia. The Holland Land Company held title to most of the land west of the Genesee River and would obviously benefit from a canal through its territory. James Geddes, who later worked on many other canals, ran the visual surveys and found the crucial level across the Irondequoit Valley near the Genesee River.

We may share his sense of exhilaration as he carried the new canal technology into the western New York frontier. He wrote later of his discovery of a series of ridges that would carry a canal "in many places of just sufficient height and width for its support," and he envisioned an embankment that would later be added. "I had, to be sure," he wrote, "lively presentiments, that . . . boats would one day pass along on the tops of these fantastic ridges, but that for myself, I had been born many, many years too soon. There are those, sir, who can realize my feelings . . . if I felt disposed to exclaim *Eureka*, on making this discovery."[7]

Gallatin's report of 1808 recommended a canal to Lake Ontario and another around Niagara Falls, instead of an overland canal, but it held out the prospect of national aid. But when New Yorkers went to Washington for help, national assistance was not forthcoming. Nor did visible support follow the speech in 1810 in Congress by Peter B. Porter of Black Rock on the Niagara River urging the binding of East and West by national internal improvements, including a canal to Lake Erie on the Ontario route.

Instead, the Erie Canal project moved forward under state sponsorship. Renewed petitions from the Western Inland Lock Navigation Company led the New York legislature to authorize new surveys, to be carried out by a board of canal commissioners, whose work would furnish a precedent for almost every other state-sponsored canal in America. De Witt Clinton entered the New York canal movement as a leading Republican member of the canal commission, which included the Federalist Gouverneur Morris and the anti-Clintonian Republican Peter B. Porter. The commissioners were appointed in 1810 to examine both routes, one overland to Lake Erie and the other via Lake Ontario.

Their report, written in 1811 by the ebullient Gouverneur Morris, showed their self-generated activity. Two commissioners sought land cessions from the great western New York landholders; two began the search for an engineer; and Clinton and Morris were soon in Washington to request congressional funding. They returned empty-handed, but another report of the commissioners in 1812 carried the movement forward in spite of war with England and military operations on the western New York frontier.

The reports of these commissioners repeated already familiar arguments over the growth of western New York, rival routes, and the need for state or national help in the canal's construction. As Darwin H. Stapleton has written, the commissioners had Philadelphia connections, and they also had an eye to movements in Pennsylvania to draw the trade of the Susquehanna, Allegheny, and Ohio valleys to Philadelphia.[8] Thomas Eddy and Robert Fulton tried unsuccessfully to interest Benjamin Latrobe, who had already surveyed a canal on the lower Susquehanna, in making the surveys in New York.

But opposition and obstruction confronted the commissioners. Sectional opponents along the lower Hudson River and in the southern tier of counties feared western growth or competition for which they might be taxed. The Holland Land Company vacillated in its support for a western canal. In 1810 Ellicott concluded that "Montreal will be our market." Then the company offered a grant of a hundred thousand acres of unsalable Allegheny lands and tried to get a canal route through Batavia, the seat of the company land office, only to rescind the grant when the commissioners held out for the canal path and additional concessions in quarrying and wood cutting rights. The legislature of 1816 met the rising movement for a canal by delaying taking any action. Governor Daniel Tompkins endorsed the reports of the commissioners, but he did not want to alienate Clinton's opponents. Clinton called him "the insidious enemy of the Erie Canal."[9] The seven-man board of commissioners pushed the project forward in spite of three-way political divisions involving Clintonians, Federalists such as Morris and Stephen Van Rensselaer, and Republicans such as Porter who were opposed to Clinton.

Paradoxically, construction of the Erie Canal on the overland route to Lake Erie was rushed into law as a state project under the rhetoric, if not the reality, of post–War of 1812 nationalism. Though this has been seen as the beginning of an era of state power, the language of canal advocacy in New York was nationalistic, drawing upon a continuing faith in republicanism.[10] De Witt Clinton made this nationalism and republicanism the dominant chord in a long memorial to the legislature, which he wrote for New York City and which ultimately bore more than a hundred thousand signatures. To be sure, Clinton, William Bayard, John Pintard, and other New York City canal proponents were trying to win support for the overland route, and they sought to show the special advantages to their state should New York need to build a canal alone. But Clinton wrote that New York was uniquely "both Atlantic and Western" and that the canal would provide a lasting "cement" to the Union. This had become familiar language in the canal proposals of Pennsylvania and Virginia, but Clinton feared a "dissolution of the union" if such an overland canal were not constructed. "It remains," he concluded, "for a free state to create a new era in history, and to erect a work more stupendous, more magnificent, and more beneficial than has hitherto been achieved by the human race."[11] Already the Erie Canal was coming to be known as the Great Western Canal.

In the spring of 1816, the Erie Canal project gained a momentum that was probably unstoppable. But the best that could be achieved in the legislature was a law authorizing further surveys, and the action turned more directly on events in Washington, where debate over Calhoun's Bonus Bill put national aid in the balance. If Madison's veto foreclosed national assistance, it also strengthened the pressures for New York to build the canal on her own. Moreover, the canal benefited from its attachment to De Witt Clinton's gubernatorial campaign. In 1817 the surveys were again favorable, and the delicate business of devising an expedient base for the Canal Fund was resolved. Perhaps most crucial, in the state senate Martin Van Buren abandoned his opposition to construction of the canal. Although he had long opposed Clinton in New York politics, Van Buren now backed the canal and added a significant element to the canal package. It would not be financed by the Canal Fund alone but would be build on the credit of the state.[12] In April the canal bill became law, and the following month its success helped elect Clinton governor, a position from which he would begin the project he had made his own, the Erie Canal.

Construction began at Rome on July 4, 1817, in a ceremony marked by a nationalistic, republican symbolism that would be emulated again and again on other canals. Political and technological considerations blended as construction began on the easier middle section of the canal between the Mohawk and Seneca rivers. With work on this section under way, the legislature would be compelled to support the canal and the eastern and western sections, where the engineering challenges were most difficult. Even with the surprising success of construction on the first ninety-two miles, political opposition continued, and completion of the entire canal was by no means assured.

The southern tier of counties had always opposed the canal because they did not expect to benefit from it. Moreover, Peter B. Porter, who had run against Clinton for governor in 1817, still wanted the Ontario route instead of a western section because it would mean a canal around Niagara Falls through Porter's lands. In addition, the canal became immersed in the personal politics surrounding the mercurial career of De Witt Clinton. Clinton, and to some extent the issue of whether the canal should be halted at the Seneca River in favor of the Ontario route, aroused the rancor of his opponents in Albany and especially in New York City.

The shrillest voice raised against Clinton and the Erie Canal was that of Mordecai Noah, editor of the *New York National Advocate* and spokesman for the Tammany Hall opposition to Clinton. Noah fulminated that the canal would become a "monument of weakness and folly." To be sure, Clinton enjoyed the support of the elite of New York's political and cultural society, men such as William Bayard, Matthew Clarkson, Cadwallader Colden, Thomas Eddy, John Murray, Jr., John Pintard, Stephen Van Rensselaer, Samuel Latham Mitchell, David Hosack, and Philip Hone, most of whom were or had been Federalists.[13]

They saw the canal as an agent of commercial development, a force for moral improvement, and an instrument in the rise of New York State. Yet Clinton's political personality threatened the republican ideals espoused by his supporters. He worked a vindictive revenge upon the "Martling Men," Republicans who had opposed him, and appeared haughty, arrogant, and corrupt. He could not sustain the consensus of support that had won him the governorship and helped to win legislative support for the Erie Canal.

Against Clinton, a Republican opposition coalesced into a party called Bucktails from the furry decoration of Tammany hats, and at their head stood Martin Van Buren. In the legislative session of 1819, the Bucktails followed Van Buren's dextrous shift to support completion of the Erie Canal, while, ironically, Clinton's support shrank to a personal following and the Federalists whom he increasingly appointed to office. Nonetheless, when Clinton ran for reelection as governor in 1820, his followers took as their slogan "The Canal in Danger," and the *Rochester Telegraph* warned that Clinton's opponents were "secretly plodding [sic] the destruction of one of the greatest projects which has ever been devised."[14] Clinton was reelected, but the Bucktails took a majority in the legislature and quickly moved to demonstrate their commitment to the completion of the Erie Canal. Their appropriations would see the canal to completion but with its Clintonian sponsorship effectively removed.

With his political fortunes ebbing, Clinton did not seek reelection in 1822. But when his opponents took the audacious step of removing him from the Canal Board, they virtually assured his succession to the governorship again in 1824. He used his martydom and the support of the new People's party to return to power just in time to preside over the completion of the Erie Canal. Political conflicts in New York, always byzantine, entangled the canal in the personal ambitions of De Witt Clinton and the shifting response of Martin Van Buren and his followers. But the growth of popular enthusiasm for the canal and Bucktail maneuvering allowed completion of both the western and eastern sections, and the new state constitution, written in 1821, guaranteed the Canal Fund.[15] The emerging canal technology virtually invited a canal between the Hudson and Lake Erie on a route favored by nature and obvious in commercial benefits. The political process required to bring this great project to fruition, however, took almost twenty years and often turned on the personal ambition or opportunism of party maneuvering before the construction of the Erie Canal could be finally completed.

The forty-by-four-foot dimensions of the Erie Canal and its ninety-by-fifteen-foot locks were modeled after the Middlesex Canal and once proven successful became almost standard for the Canal Era. Construction was directed by three acting commissioners, one on each division, and canal finances were managed by the commissioners of the Canal Fund. Again New York created the administrative model to be copied by other states. Benjamin Wright was chief engineer, assisted by James Geddes on the middle section. Other engineers, including David Thomas, Nathan S. Roberts, David S. Bates, and John B. Jervis, rose through the ranks as they were trained during their work on the canal. As Charles G. Haines, Clinton's private secretary, wrote in 1821, "The canal line is now one of the most excellent schools that could be divised [sic] to accomplish men for this pursuit."[16] Before the Erie Canal, no other state or region had produced a distinctive source of engineers. There were the Baldwin brothers in Massachusetts, William Strickland and Robert Mills, who had trained under Benjamin Latrobe, and army engineers who had trained with Claude Crozet at West Point.[17]

The accomplishments of these engineers are the more remarkable because the state had no public works apparatus or work force. Contracts were let locally, usually for a mile or more to contractors who hired their own labor and were in effect each excavating a tiny canal. More than fifty contractors worked on the first fifty-eight miles, following a set of specifications applicable to every contract. Contractors were paid at the rate of ten to fourteen cents per cubic yard for the excavation of earth, seventy-five cents for marl, and as high as $2 for breccia, a particularly hard rock. The principal engineers received a salary of $1,500 to $2,000 a year.[18]

The tradition that Irish workers were recruited abroad to dig the canal requires qualification. The New York canal commissioners reported in 1819 that three-fourths of the laborers were "born among us." But after 1821 Irish and Welsh immigrants appeared in greater numbers. The depression of 1819 made surplus labor available, and wages were low—$8 to $12 a month or fifty cents a day. A letter in a Rochester newspaper complained in 1820 that the only resource for wages was "by getting a job on the canal." By 1819 between two and three thousand men were at work on the middle section of the canal, which was completed in three years.[19] On this section, one of the "long levels" ran sixty miles without a lock between Rome and Utica, pitched at a drop of one inch to the mile to provide a gentle current of water. On July 4, 1820, seventy-three boats passed through the canal to open the middle section.

But troubles multiplied and some of the glow of progress diminished. The commissioners' report for 1819 accounted about a thousand men sick from heat and the necessity of working in wet soil between Syracuse and the Seneca River in the months of July to October, which, they noted, was "a most discouraging spectacle." Their report for 1821 noted continued sickness on the level west of the Seneca River, which began in August and became so widespread that among the two to seven hundred men at work, "all the principal contractors, with many of the sub-contractors and hands, became diseased." Meanwhile, heavy rains "kept the earth almost constantly muddy, heavy and slippery, insomuch that every exertion was made to great disadvantage." Such conditions caused delays, but the commissioners assured the legislature of "our entire conviction, that our plan of making the canal was practicable."[20]

Their plan proved practicable partly because innovative techniques emerged in almost every aspect of canal building. The plow and scraper were used more often than the spade. Jeremiah Brainard of Rome designed a wheelbarrow made of a single board bent to a semicircular shape that was easier to unload than earlier models. Trees were brought down using a cable attached a hundred feet up their trunks, which wound around a roller turned by an endless screw and crank. A giant stump puller used a pair of wheels sixteen feet high, connected by a thirty-foot axle into which a fourteen-foot wheel was spoked and around which a rope was wound several times to produce an eightfold gain in power for grubbing out a stump. A suitable clay was found for puddling or lining the canal banks to keep them from leaking.

The perfection of an underwater cement was the invaluable achievement of Canvass White of Whitesboro, an engineer who began as a farm boy and was trained to assist Benjamin Wright with the surveys. He was sent to England, where he examined the underwater cements used there. He returned to find a variety of limestone that could be made into a quicklime cement superior to any made in America. More than four hundred thousand bushels were used on the Erie Canal. White took out a patent on the cement, and his factory supplied other canals in the Canal Era. When in 1878 the largest lock built to that time was made part of the

Davis Island Lock and Dam on the Ohio River at Horse Tail Ripple below Pittsburgh, War Department engineers sent to Canvass White's factory for cement.[21]

Each of the eighty-three locks on the Erie Canal was built with cut stone, floored with foot-thick timbers over which were placed two layers of planks. Mitered lock gates bound in iron were fitted with ingenious sluices and wickets for releasing the water from a lock while the gates remained closed, and locks were opened or closed by long herringboned wooden arms balanced to be operated by a single man. At Lockport on the Niagara escarpment the five-tiered locks placed side by side in twin sets carried the canal up and down a rise of sixty-six feet like a staircase. They were designed by Nathan Roberts and David S. Bates. Paintings or drawings of Lockport at the head of a high-bluffed gorge, mounted by this five-lock combine, have been selected most frequently to illustrate work on the Erie Canal or any work on American economic development during this period.

But the eighteen stately aqueducts were the glories of the Erie Canal. These great stone structures, which combined beauty and great strength, rested on piers set in the rivers they crossed. The piers were joined by beautiful Romanesque arches, expressions of the classical revival architecture of the period. The aqueducts carried a wooden trough through which the boats could pass, while above the trough a high towpath was made just wide enough for the boat-pulling horses to cross.

Two majestic aqueducts crossed and recrossed the Mohawk River below Schenectady in the Mohawk Valley. At Alexander's Mills (later Rexford) the aqueduct that carried the canal to the north side of the Mohawk was 748 feet long and rested on sixteen piers. Twelve miles below, the canal was brought back to the southern side by the longest aqueduct that stretched almost 1,200 feet and rested on twenty-six piers. At Little Falls a navigable feeder crossed the Mohawk on a single graceful arch with a span of 70 feet. The Rochester aqueduct that crossed the Genesee River was a stone structure 802 feet in length, resting on eleven solid, squat arches. Its wooden trough was an unusual 45 feet wide so that boats could pass on the aqueduct. But for sheer charm, perhaps nothing could surpass the small aqueduct at Palmyra, the location for Samuel Hopkins Adams's 1945 novel *Canal Town*.

Nearby at Irondequoit, the embankment envisioned by James Geddes in 1807 became a reality in 1822 and allowed water from the

Genesee River to flow east in the canal. Here was a daring triumph, an earthen mound making a level seventy feet above the valley floor and covering a culvert over Irondequoit Creek. West of Rochester, in the opposite direction, another long level ran sixty-four miles to Lockport, straight and classical as a Roman road. Hydraulic weighlocks were devised, the one at Rochester resembling a Greek temple. It still stands and serves as a canal museum. Romanesque and Greek revival structures and canal towns with names such as Rome, Utica, and Syracuse gave testimony to the celebration of classical republicanism.[22]

Impressive as such engineering achievements were, two termini of the canal presented the greatest challenges. At the western extremity the canal ascended the Niagara escarpment with the five-lock Lockport combine, but the level would also require seven miles of deep cutting, including two miles through solid rock, to reach the Niagara River. Until 1825 it was still undecided whether the canal would terminate at Black Rock in a harbor in the swift current of the Niagara River as desired by Peter B. Porter or at Buffalo as desired by James Wilkeson, where a sandbar blocked the entrance of lake boats into Buffalo Creek. Wilkeson and the Buffalonians with heroic effort created a harbor, and the rivalry with Black Rock on the Niagara River for the canal terminus was both frenzied and comic. The question was resolved by building two harbors, at Black Rock and Buffalo, connected by a stretch of canal parallel to the Niagara River. When the Black Rock harbor failed, Buffalo on Lake Erie became the terminus. De Witt Clinton sided with the Buffalonians, and his comment on the failure of the Black Rock harbor suggested the importance of the controversy: "Conceived in sin, fed by ignorance and brought forth in iniquity, it had disappointed no observing man—as long as the Canal lasts its ruins and history will be subjects of reproach to the projectors."[23]

Similar difficulties faced canal construction in the lower Mohawk Valley. "How we shall get a line from Schenectady to the Hudson I am most anxious to know," wrote a canal commissioner to Clinton.[24] The problem was to pass Cohoes Falls below Schenectady. That obstacle was overcome by a line of twenty-seven locks in thirty miles reaching the Hudson River near Watervliet and then by directing the canal five miles along the Hudson south to Albany. In a sense there were several termini at the eastern end of the Erie Canal. For most passengers, the canal ended at Schenectady above Cohoes. From there they took the stage or the Mohawk and Hudson Railroad to Albany and saved a day's canal travel. Freight boats passing down the locks to the Hudson level passed through Juncta, where the Champlain Canal connected to Lake George and Lake Champlain, and entered the Hudson at West Troy or Watervliet. But some boats crossed the Hudson River to Troy, which received about a third of the commerce of the canal. And most boats continued south past the United States Arsenal five miles further to the Albany Basin, which was the major eastern terminus of the Erie Canal.[25]

Controversy over this plan was less frenzied than that for the western terminus, but it became public when one of the engineers, John Randel, Jr., published a pamphlet challenging Benjamin Wright's decisions. Randel's plan need not be described here, but it is noteworthy as a precursor to the far greater conflict between Randel and Wright on the Chesapeake and Delaware Canal, which involved that waterway in a long and expensive litigation. And these two New York engineers were again locked in controversy on the James River and Kanawha Canal in Virginia.

By 1823 construction had progressed rapidly enough to allow navigation from Brockport, just below the Niagara escarpment, all the way east to Albany. Even before the last western segment to Buffalo was completed, the canal had earned $1 million in tolls, and its burgeoning commerce foretold its phenomenal success once the entire waterway was completed in 1825. Measured against such earnings, the total cost of $7,143,789 seemed small, and the canal's potential returns appeared almost boundless.[26] When the celebration marking the completion of the Erie Canal took place in October and November of 1825, two thousand boats, nine thousand horses, and eight thousand men were already employed on the canal.[27] And when the Marquis de Lafayette traveled down the nearly finished canal in the spring of 1825 as part of his great circle tour of the United States, his journey was a preliminary celebration of the Erie Canal.

The celebration of the completion of the Erie Canal as the "Wedding of the Waters," the mixing of the waters of Lake Erie and the Atlantic, was one of the most spectacular events of the early national era. It struck the consciousness of Americans in much the same manner as would the driving of the golden spike at Promontory in Utah on the Union Pacific Railroad in 1869 and the celebration in New York Harbor of the centennial of the Statue of Liberty in 1986. As Page Smith has written, "The historian has . . . difficulty . . . in suggesting the degree to which the canal obsessed and enchanted Americans in the fall of 1825. . . . It was taken to be a symbol of the boundless potentialities of the country, its resiliency and its hopes."[28]

The celebration began at Buffalo on Lake Erie on October 26 and rose to a crescendo in New York City over the next ten days. The *Seneca Chief*, with Governor Clinton, the canal commissioners, and the engineers on board, led a slow-moving procession down the canal to Albany, stopping for speeches and festivities all along the way. Also on board the *Seneca Chief* were two kegs of Lake Erie water, which were to be poured into the Atlantic off Sandy Hook. As the procession left Buffalo, the first thirty-two-pounder of the "Grand Salute" was fired, telegraphing the news of the departure by guns spaced earshot distance apart all the way to New York City. Three hours and twenty minutes later the Buffalonians heard the rumbles from the East signaling that the salute had been received and returned. The exercise had cost the lives of two gunners at Weedsport killed when their cannon discharged. The ever-lengthening procession of boats passed down the canal beneath garlanded arches as immense crowds gathered along the banks. The towpath was crowded with horses and carriages while bands played and cannon sounded. Arriving at Albany on November 2, the procession met the grandest celebration of all on the canal, complete with a parade, a banquet, and speeches at the capitol. Then eight steamboats guided the official party down the Hudson to New York Harbor for the "Wedding of the Waters" on November 4.

From New York Harbor the "Grand Aquatic Display" led out to the Atlantic off Sandy Hook. A fleet of forty-six vessels, dressed out in flags and bunting and with bands blaring, dwarfed the four tiny canal boats towed along in their midst. Finally, from the deck of the United States schooner *Porpoise*, Clinton poured a keg of Erie water into the Atlantic. Awaiting their return at the Battery was the "Grand Procession" in readiness for the parade to City Hall. The *Memoir . . . at the Celebration of the Completion of the New York Canals*, put together for the occasion by Cadwallader D. Colden, describes the fifty-nine units containing some five thousand marchers lined up behind the standard of their profession, trade, or society.[29] Twenty trades were represented, and many pulled cars on which they demonstrated their crafts and revealed much about their status relationships as well. Speeches, banquets, fireworks, and a Grand Canal Ball lasted on into the night. When it was over, the *Seneca Chief* made its way back through the canal, this time carrying a keg of Atlantic water to be mixed with the waters of Lake Erie.

The sheer magnitude of New York's achievement in the Erie Canal evoked a surfeit of speeches and newspaper editorials and the lavish accounts by Cadwallader Colden and William Stone, which were bound together in Colden's *Memoir.* They were a celebration of triumphant republicanism persisting from the more experimental years of the Federalists and the Jeffersonians and now extended in an expanding, westward-moving nation. Clinton's speech to the deputation that met him on the Hudson to welcome him to New York City emphasized the importance of the canal to the "duration of the Union" and the "Holy cause of Republican Government." Colden wrote in his *Memoir* that the ceremonies of 1825 were held partly to attract the attention of foreign nations: "They have told us that our government was unstable," he wrote, "that it was too weak to unite so large a territory—that our Republic was incapable of works of great magnitude . . . but we say to them, see this great link in the chain of our union—it has been devised, planned and executed by the free citizens of this Republican state."[30] The eulogies to De Witt Clinton strengthened his partisan triumph over the Bucktails, and there was never a doubt that New Yorkers welcomed their new waterway as a cornucopia of commerce that would enrich their coffers and promote the development of their state.

During the first decade of canal commerce, most products came from western or central New York. Little entered the canal at Buffalo. Paul Wallace Gates has written that New York was in a "Golden Age of Agriculture" by midcentury, exceeding all other states in the number of farms, value of farms, number of dairy cows, and production of butter, cheese, milk, hay, potatoes, hops, maple sugar, and especially wheat. The wheat district reached westward for 150 miles from Cayuga County to Lake Erie, and wheat and flour were first in value on any list of products shipped to tidewater until about 1844.[31] By 1832, in the canal's first seven years, almost half a million barrels of flour and 146,000 bushels of wheat were deposited at Albany. Four years later, these figures were almost doubled, reflecting the rise of Great Lakes commerce entering the canal at Buffalo.[32] Although the grain trade dominated Erie Canal commerce, forest products were second in volume, filling the stream of canal boats heading toward tidewater. Some 36 million feet of timber and scantling went east on the canal in 1832.[33]

The number of canal boats carrying this commerce jumped from the 14,300 that arrived at Albany in 1832 to 18,850 in 1834; the latter year they carried 156,000 tons.[34] The increase of western commerce was so rapid, and so much more seemed certain to come, that New York began a program to enlarge the canal in 1835 to channel dimensions of 70 feet by 7 feet with locks 110 feet by 18 feet. Some double locks were also authorized, which allowed boats three times larger than those of the original canal, and they could pass twice as quickly. In effect, the Erie Canal was rebuilt, for its greater depth allowed straighter paths and longer levels. Reconstruction began in 1836 on the eastern section, where the press of boats was greatest, and, amazingly, was carried on while the original canal continued to operate. Such repeated construction while navigation was maintained became typical of the Canal Era.

The Panic of 1837 brought a financial crisis which slowed canal construction in almost every state that built canals. In New York a "stop and tax" law in 1842 suspended expenditures on public works and nearly brought construction on the Erie Canal enlargement to a halt. The law imposed a tax to begin payment on the state debt. Surplus revenues allowed resumption of construction at the end of the decade, but the enlargement was not completed until 1862. It is therefore often necessary to specify which Erie Canal, where, and at what time.[35]

Meanwhile, in addition to rebuilding the Erie Canal, New York State added lateral canals. An act known as the "great canal law" of 1825 authorized surveys for seventeen new canals. Some of those selected for construction added significantly to the commerce of the Erie Canal and aided the development of the state. Others were largely responses to local interests, had little traffic, and remained a drain on the New York canal system. The most successful was the Oswego Canal, which followed the Oswego River from Syracuse to Oswego on Lake Ontario and was completed in 1828. In the Finger Lakes region, the Cayuga and Seneca Canal was completed in 1829; the Chemung Canal, extending from Seneca Lake to Elmira and supplied by a feeder from the Chemung River, was finished in 1833; and Seneca Lake was linked with Keuka Lake in the same year. The Erie Canal was joined with Oneida Lake by five miles of canal in 1835 and with the Susquehanna River through the Chenango Canal in 1837. The Chemung and Chenango canals became important links to the Pennsylvania canals reaching northward along the upper branches of the Susquehanna.

In a dramatic feat of canal engineering, the Black River Canal was constructed from Rome to the Black River in the foothills of the Adirondacks. Successful only in its engineering, this canal was finished in 1851. Western New York won its connection by canal to the Allegheny River through the Genesee Valley Canal, built from Olean on the Allegheny to Rochester at the falls of the Genesee River. Construction on this ill-starred lateral canal stopped at Mt. Morris in 1841, and the canal did not reach Olean until 1856. Almost always New Yorkers contested with each other as to which way the laterals should go, much as they had contested the termini of the Erie Canal at Black Rock or Buffalo, at Troy or Albany.

With the growth of western settlement, the larger dimensions of the canal, and the use of larger boats, canal commerce increased dramatically. By 1850 the Erie Canal carried almost 23 million bushels of wheat and flour, a quarter of all the grain produced in the nation. In 1851 the volume of western goods, chiefly agricultural and forest products, exceeded a million tons. By 1856 8.5 million bushels of wheat were sent east from Buffalo. Between 1836 and 1856 the volume of products from western states increased by 500 percent while those from New York grew by 80 percent.[36]

Freight rates fell and freight charges (without tolls) were $4.16 per ton eastbound and $9.00 per ton westbound during the period 1830 to 1834. Ten years later, between 1840 and 1844, the eastbound rate was $2.96 and the westbound rate was $6.67. In the last five years before 1860, the eastbound rate dropped to $1.75 and the westbound rate to $2.28. By 1850 freight charges were reduced to approximately 1.7 cents per ton-mile.[37]

The lowering of tolls during the 1840s and 1850s encouraged a steady rise in canal tonnage. In 1837 the Erie Canal carried 667,151 tons of freight. More than a million tons were carried in 1845, and by 1852 more than 2 million tons, a figure that would be maintained throughout the 1850s. The 2,491,495 tons carried to tidewater in 1853 were valued at $74,443,061. Western products came to exceed in volume those from New York by 1847, and by 1860 almost 90 percent of the tidewater receipts came from the West, contributing 80 percent of total revenues.[38]

Competition from Canada for western trade, which had compelled the construction of the Erie Canal, continued with the construction of the Welland Canal through Canada around the Niagara River, and Canada completed a network of canals on the St. Lawrence River in 1832.

Although the Canadian system offered a series of ship canals, the American advantage held. The progress of the Erie Canal's enlargement, day and night lockages, the doubling of locks east of Syracuse, and the improvement of grain elevators at Buffalo and Rochester all doubled the capacity of the Erie Canal by 1840 and increased it by another 50 percent by 1848.[39]

Upward canal trade, largely manufactured goods destined for country stores, amounted to 560,754 tons in 1853, a far smaller volume than downward canal trade, but was valued at $114,090,801.[40] And human freight went up the canal as well. Packets carried immigrants by the thousands who turned Buffalo into a great gathering place for transfer from canal boat to schooner and steamboat, a jumping-off place much like St. Louis as a gateway to the West. An observer in 1832 wrote the following description of this transfer from canal to lake:

> Canal boats filled with emigrants, and covered with goods and furniture, are almost hourly arriving. The boats are discharged of their motley freight, and for the time being, natives of all climates and countries patrol our streets, either to gratify curiosity, purchase necessaries, or to inquire the most favorable points for their future location. Several steamboats and vessels daily depart for the far west, literally crammed with masses of living beings to people those regions. Some days, near a thousand thus depart. As I have stood upon the wharves and seen the departure of these floating taverns, with their decks piled up in huge heaps with furniture and chattels of all descriptions, and even hoisted up and hung on to the rigging; while the whole upper deck, and benches, and railing, sustained a mass of human bodies clustering all over them like a swarming hive—and to witness this spectacle year after year, for many months of the season, I have almost wondered at the amazing increase of our population, and the inexhaustible enterprise and energy of the people! What a country must the vast border of these lakes become! And Buffalo must be the great emporium, and place of transit for their products and supplies.[41]

The Erie Canal constituted a virtual river of humanity flowing across the state. In a remarkable accommodation, New Yorkers subjected the canal to system and organization, and for all the ethnic tensions that followed its impact, transients came to be viewed with welcome rather than fear. A report in a Rochester newspaper in 1825 told of those who went to the canal basin and moved close to a newly arrived boat "to gaze upon the passengers, and to learn their names, condition, destination and business."[42]

Although the primary significance of the Erie Canal rested on the freight it carried, its packet travel was more extensive than that on any other canal of the Canal Era. It took more immigrants from an Atlantic port to the Great Lakes than any other trans-Appalachian canal, adding a northern influx of settlers to the states bordering the Ohio Valley. Though it did not replace the stage and was soon superseded by the railroad, its waybills reveal local usage as well. Packets offered an attractive alternative to poor roads and an opportunity to travel day and night and eat on board, and the canal was a comparatively cheap mode of transportation.

At first, packets represented the shock of the new. A Rochester pioneer wrote of her first journey by canal, "Commending my soul to God, and asking his defence from danger, I stepped on board the canal boat, and was soon flying towards Utica." A canal boat packet company was organized in 1820 at Rome, New York, two more in 1823, and a fourth in 1824. Attempts at monopoly failed, and some smaller operators sent one or two boats plying back and forth on the canal. In 1836 four daily lines of packets left Schenectady for the West. Rochester was served by three daily departures to the east and six to the west. The journey from Albany to Buffalo normally took four to six days in the 1830s and 1840s, but fast packets could make it in fifty to seventy hours. More commonly, packets covered eighty miles in twenty-four hours. From Schenectady, where most passengers boarded the Erie Canal for Buffalo, the fare in 1846 was $7.75 with board, $5.75 without.[43]

Line boats carried both passengers and freight, offering cheaper travel but a slower passage than the packets. They commonly covered sixty miles in twenty-four hours. Passenger travel was strong into the 1840s, and packets entered a golden age of luxurious design. In 1843, the Red Bird Line, out of Brockport in western New York, had packets one hundred feet long that berthed one hundred passengers. A Rochester paper in 1842 noted packets that "run full both ways," one "out of Utica . . . with 170 on board."[44] By 1849 the fare from Schenectady to Buffalo had dropped to $6.50, including board.

The new inland boat-building industry was centered in Rochester. By 1846 eleven boatyards there turned out packets and freighters excelling any others afloat. That year they launched 210 boats with an average value of $1,300 each. Six years later some 261 men were employed in Rochester boatyards, and the boats they produced were valued at more than $300,000.[45] Packet travel, however, declined in the 1840s in the face of railroad competition.

Rivalry from the railroad began with the Mohawk and Hudson Railroad between Schenectady and Albany, which started to operate in 1831. In 1853, seven short railroad lines were consolidated into the New York Central Railroad running between Albany and Buffalo. For a time, the state refused to allow railroads to carry freight except when the canal was closed and compelled railroads to pay canal tolls until 1851. But in the 1850s, the railroad carried 66 percent of the high-valued freight and 22 percent of the low-valued freight across New York. It carried almost half of the way freight but less than a fifth of the through freight.[46] Unlike most of the canals of the Canal Era, the Erie Canal survived the rise of the railroad. Tonnage on the Erie Canal continued to grow and finally peaked in 1872.

The canal continued to provide a cornucopia of wealth for the state. The tolls of $687,976 collected in 1826 were doubled to $1,375,673 collected in 1835. In 1847, the high point of toll receipts before the Civil War was attained with collections of $3,333,347. Though tolls declined in the 1850s, by the time they were finally abolished in 1883 the Erie Canal had earned total revenues of $121,461,871.[47] The lateral canals in New York contributed to the earnings of the Erie Canal, and both the Champlain and Oswego proved profitable. Other laterals such as the Chenango, Genesee Valley, and Black River canals, however, yielded only fractions of their total cost.

For the remainder of the Canal Era, the success of the Erie Canal was a stimulus and a challenge to other states that sought to imitate the Erie Canal or outdo New York. Adjacent to New York, New England lay between Albany and the Atlantic, and its rivers ran parallel to the Hudson toward the sea. Moreover, many of New York's canal builders were from New England, and they would return to build canals in the region of their birth.

Fortunately, it would seem, that region delayed building an east-west canal link from Boston to Albany, which the mountains would have doomed from the start. When the link was made, it was by rail. Meanwhile, the fast-running rivers flowing toward Long Island Sound would be adapted to more navigable canals. In his classic history of New England transportation, Edward C. Kirkland described a "second canal era" in the 1820s after the earlier improvements by the Middlesex Canal and the canals along the Merrimack. Kirkland wrote that the Erie Canal "stirred and quickened the zeal of New England communities ambitious for commercial greatness. Almost in a twinkling the whole region was criss-crossed with a network of canals—on paper."[48]

The Middlesex Canal brought the trade of the Merrimack down to Boston from the north; it prospered in the 1830s and until 1852. Inland from Boston, the area around Worcester lay below a watershed stretching north and west, and the Blackstone River flowed southeast to Narragansett Bay toward Providence in Rhode Island. For the Browns, the Iveses, and other great merchants of Providence, the Blackstone Valley presented an undeveloped pathway to central New England. The Blackstone Canal became the outlet for the commerce of Worcester and the hope of Providence to profit from the development of the interior.[49]

John Brown and others in Providence had proposed such a canal to Worcester in 1796 in a project similar to that of James Sullivan and Loammi Baldwin in the Middlesex Canal and of Henry Knox for a canal from Boston to the Connecticut River. The *Massachusetts Spy* described the project as a canal "into the heart of the county of Worcester . . . for the purpose of its being carried into and through the county of Worcester to some part of Connecticut River." The Rhode Island legislature incorporated the Proprietors of the Providence Plantations Canal to build a canal "toward the Town of Worcester," but no action was taken.[50]

The Blackstone Canal Company was incorporated in both Rhode Island and Massachusetts in 1823. It was backed by a financial elite much like the entrepreneurs then organizing a canal company in New Haven to gain the trade of the Connecticut Valley. The Blackstone Canal Company, which set out to build a canal along the forty-five-mile route through the Blackstone Valley, included among its proprietors Nicholas Brown, Thomas P. Ives, Edward Carrington, Moses Brown Ives, Cyrus Butler, and Sullivan Dorr. Stock was oversubscribed almost immediately, and the canal was completed in 1828.[51]

Construction began in 1824 when the company hired Benjamin Wright as engineer and Holmes Hutchinson as assistant engineer. Vincent Edward Powers has described the canal as the "step-child of the 'big ditch' of New York." Benjamin Wright brought contractors and crews from the Erie Canal, but only five miles were done in 1824. The completion of the Erie Canal in 1825 made more labor available, and in 1826 a thousand men were at work, over 80 percent of them Catholic Irish. By the end of the year the canal reached from Providence to Mendon in Massachusetts. Meanwhile, construction moved south from

Worcester, covering the twenty-three miles to Uxbridge. There the contractor was Tobias Boland, who brought five hundred laborers to work on this difficult section. Forty-eight granite locks were built to meet the 451 feet of elevation on the whole canal line, and its cost rose to $550,000, an amount that would never be returned to the investors.[52]

Powers has studied the Irish laborers who built the Blackstone Canal and has followed their lives in an "Irish colony" in Worcester. There wages ranged from twenty-seven to forty-five cents a day. Many of these workers were sons of "middling class Hibernians." Many had special skills, and some acted as supervisors. Powers found that they had worked on the Erie Canal, were drawn by kinship ties from Ireland, and had previously worked on waterways throughout the British Isles.[53]

When additional funds were needed after the canal was completed in 1828, Nicholas Brown and Thomas P. Ives subscribed the largest share of the new loan certificates that were issued. An effort to interest Prime, Ward, King, and Company in New York was unsuccessful, and the company turned to an improvement bank for new capital. The Blackstone Canal bank purchased new stock, and the company paid small dividends in 1832, 1834, and 1835.

The city of Worcester became a canal port. Land increased in value, and shipping costs from Worcester to Providence dropped from $6 to $1 a ton. In its heydey in the 1830s the Blackstone Canal seemed as successful as the Middlesex Canal to the north. Twenty or more boats, along with a luxurious packet, carried trade in flour, salt, molasses, lumber, and coal. Tolls rose to $18,907 in 1832 and then declined to $11,520 in 1836. The usual predominance of down-canal traffic was reversed as tonnage in 1834 was 12,761 up the canal and only 4,743 down.[55] More important, the canal tied Worcester to the Rhode Island port of Providence rather than to Boston or west to Springfield.

But the Blackstone River textile factories found the canal a competitor for the water that powered their mills, and they fought the canal, which benefited chiefly the cities at either end. Perhaps for the canal to have succeeded, it should have been built like the Middlesex Canal in the 1790s, before the Blackstone mills had been founded and the conflict over riparian rights developed. Moreover, new competitors which none could have foreseen appeared in two railroads, one from Boston to Worcester in 1835 and one from Providence to Worcester in 1847. Just twenty years after it had been completed, the Blackstone Canal ceased operations in 1848.

Yet the Brown and Ives families maintained their cotton mill interests and their banking activities and remained committed to the Blackstone Canal until the end. James B. Hedges suggested the importance they placed on the project when he wrote that "they helped launch an enterprise foredoomed to failure but one on which they staked a considerable capital, great energy, and genuine devotion."[56] Such devotion was exhibited over and over again by the promoters who had great faith in canal projects that were later regarded as endeavors in futility or doomed from the start.

The most promising canal in all of New England was understandably tied to the great river of that beautiful region, the majestic Connecticut. Its valley was rich in agriculture and dotted with mill towns, where water power turned the wheels of manufactures. The periodic river obstructions invited speculation that improvements could be made by building canals. As early as the 1790's the South Hadley Inclined Plane had pointed the way toward such improvements and Loammi Baldwin, Jr., had proposed a series of slackwater navigation projects.[57]

From the beginning there was rivalry over what kind of canal and what route would best serve the valley. Partisans of Hartford, where unimpeded river navigation stopped, favored river improvements almost exclusively in the upper valley, with dams or canals built at obstructions such as those at Enfield Falls. They found precedents in earlier efforts to surmount the steep river descents at South Hadley, Miller's Falls, Bellows Falls, Queechee Falls, and Olcott's Falls below Hanover. Hartford interests chartered the Connecticut River Company in 1824 to improve the entire northern river, and they hired Holmes Hutchinson, the Erie Canal engineer, to design river improvements estimated to cost $1.5 million. In 1829 the company completed the Enfield Canal, six miles long with three locks and an aqueduct. Canvass White from New York was the engineer, and the short canal boded well for future river improvements.[58]

Meanwhile, New Haven launched the project that would become the New Haven and Northampton Canal but turned out, as George Rogers Taylor described it, to be "the longest, most costly, and least successful of the New England canals."[59] New Haven was some distance from the mouth of the Connecticut River so the city developed a canal plan that would tap the river at Northampton and draw its waters over a seventy-eight-mile route to New Haven, bypassing Hartford. James Hillhouse, treasurer of Yale University, was the moving spirit behind the plan, and

Simeon Baldwin carried it forward. A group in Farmington, envisioning a connection through the Farmington River Valley to the Erie Canal, petitioned the Connecticut legislature, which quickly chartered the Farmington Canal Company in 1822. Simeon Baldwin headed a six-man commission to construct a canal from New Haven to the Connecticut border at Southwick. The following year, Massachusetts chartered the Hampshire and Hampden Canal Company to continue the canal to the Connecticut River.[60]

The Farmington Canal Company hired Benjamin Wright as engineer. He employed Eli Whitney Blake to take the levels, using a leveling instrument made by his uncle, Eli Whitney. Wright was assisted by his son Henry, and he also engaged the engineer brothers Davis and Jarvis Hurd. Jarvis Hurd soon became chief engineer of the Connecticut part of the line, the Farmington Canal. Surprisingly, Wright recommended wooden locks, which soon rotted, rather than stone, probably because of a shortage of funds.

Construction began at Southwick on July 4, 1825, the year of the completion of the Erie Canal and the day De Witt Clinton turned the first spadeful of earth on the Ohio canals. Governor Oliver Wolcott turned the first spadeful of earth at Southwick, and in retrospect it was taken as a bad omen that the handle of the spade gave way. The fifty-eight miles of the Farmington Canal were completed in 1828 though the twenty-mile Massachusetts part of the line had barely begun.

On the Hampshire and Hampden Canal in Massachusetts, Davis Hurd was chief engineer, and the work was let in only two contracts. One beginning at Northampton was let to Thomas Shepard, who built much of the canal with his personal funds until he lost some $75,000 by 1832. He regained his losses in later business successes.[61] The other contract on the Hampshire and Hampden was let to Jarvis Hurd and Thomas Sheldon of Westfield. The Hurds left the canal in 1829 and Sheldon finished the line from Southwick at the state line north through Westfield to the Little River below Southampton. There was heavy canal traffic on the Farmington Canal north as far as Westfield in 1833 and 1834, and in 1835 the entire line from New Haven to Northampton was completed. When the first boat through New Haven was met by one from Northampton, George Bancroft stood on the deck of the latter to give the welcoming speech. His famous history of the United States had just been published. Absent was James Hillhouse, the canal's founder and president of the Farmington Canal Company, who died in 1832. The completion of the Massachusetts line had lagged five years behind that in Connecticut, and the total cost of the canal was $2 million.[62]

By the time the canal was completed, the heavy cost of construction and repairs had exhausted the original companies. Eight assessments were made on the stock of the Farmington Canal Company, and the Mechanics Bank in New Haven was chartered to finance the canal. Thomas Sheldon completed the Massachusetts line to Northampton. He was considered a rich man, the most prominent in Westfield, whose standing was achieved by his work on the canal. But his fortunes declined with those of the canal. He left for Texas in 1837 and died the following year. In 1836 the canal became the property of the New Haven and Northampton Canal Company, and the original investors lost more than a million dollars.[63]

This canal along the Connecticut River was part of an unrealized plan for an international waterway from Long Island Sound to the St. Lawrence River, as well as a connection to the Erie Canal in New York. Efforts were under way to extend the canal north to Barnet, Vermont, and to the St. Francis Valley, where Canadian financiers would extend it to the St. Lawrence. De Witt Clinton came from New York in 1827 and pronounced the plan feasible, but the continuing opposition of the Hartford "riverites" helped block the northward extension of the canal.[64]

But with the entire canal finished and a new company organized in 1836, the New Haven and Northampton Canal promised to become the great canal of western New England, beginning in the Berkshires and descending to the sea. And so it appeared to be. In 1838 packets left Northampton for New Haven twice a day, making a trip in twenty-four hours for $3.75.[65] Canal basins became part of valley commerce, and companies advertised "canal navigation" along with their goods. The New Haven canal trade was served by 116 boats in 1839, which carried almost 4 million pounds in a long list of products.[66] In 1841 the canal carried twenty-one thousand tons and began to divert traffic from the Connecticut River. The best year for the canal was 1844, when its boats carried twenty-four thousand tons of goods.

The improvement in the fortunes of the canal in the 1840s was chiefly owing to Joseph Sheffield, who became a director of the company in 1841 and president in 1843. Sheffield was a wealthy southerner who brought his son to Yale and stayed to invest heavily in canal stock. He restored the canal to operating condition and was responsible for its best years. Henry Farnum, who had been with both parts of the canal since 1832, joined him in the management of the company. Facing the prospect of railroad competition, however, the company secured legislative approval to build a railroad on the towpath, believing that canal and railroad could operate together. While the New Haven and Collinsville Railroad was being built on the towpath, the canal continued to function.

The rise and fall of 520 feet between New Haven and Northampton, requiring sixty locks, may have meant that canal navigation could never compete successfully with river navigation, especially when steamers passed easily from Hartford to the Atlantic. The steep descent of the valley made the canal vulnerable to frequent flooding and required costly repairs, to which Henry Farnum devoted much of his life. Yet just as the canal appeared to be overcoming some of its difficulties, its owners turned their waterway into the path of a railroad, and in 1848 they closed the canal. Charles Rufus Harte, a civil engineer who has studied the New Haven and Northampton Canal closely, concluded: "It was a wisely conceived scheme to meet a real need; there was every reason to anticipate a successful and profitable outcome, and the failure to realize the expected result was caused by a combination of adverse circumstances which could not have been reasonably foreseen."[67]

The dismal failure of a canal in the populous Connecticut River Valley stands in contrast to the success of the Cumberland and Oxford Canal in Maine, constructed from Sebago Pond to Portland. The Maine legislature chartered a company to build this canal in 1821, and the route was surveyed by an Erie Canal engineer in 1825. It was only a little more than twenty miles from Lake Sebago to tidewater near Portland, and a lock in the Sango River would allow navigation forty miles further inland. Construction begun in 1828 was completed two years later, making a canal thirty-four feet wide with twenty-eight locks. The twin-masted boats used on Lake Sebago traversed this canal with masts folded to their decks.

Lumber trade and agricultural development, plus the long delay in railroad competition, gave the Cumberland and Oxford Canal a longer life span than other New England canals. It served the region northeast of Portland until the Portland and Ogdensburg Railroad went into service in 1870. Then, long after more famous New England canals had been closed, this one too ceased to operate.

New England topography decreed the local nature of these canals as they descended from the mountains toward the sea. But they were as nothing compared to the great abortive canal project that was planned to link Boston and the Hudson directly to the outlet of the Erie Canal. Early in 1825 the Massachusetts legislature established a commission to consider the feasibility of such an ambitious venture. The commission's report of 1826 was prepared by Loammi Baldwin, Jr., who had surveyed other New England canals, and it astonished its readers by its audacious proposal for a route across the mountainous northern part of the state. When the proposed canal reached Hoosac Mountain in the Bershires, with its great fall to the level of the Hudson, Baldwin would cut through the solid rock of the mountain to create a tunnel four miles long, as would be done years later by the Hoosac Railroad. Baldwin had examined canal tunnels in England and had worked on the Union Canal tunnel in Pennsylvania.[68] Such a canal might cost $6 million and would be studded with locks, aqueducts, and embankments. The commissioners expected to divert at least half of the Erie Canal trade to reach the Atlantic at Boston rather then down the Hudson to New York City. Much as railroads would do half a century later, it was projected that Boston's interior trade would extend to the Illinois and Michigan Canal then under construction. The commissioners were particularly interested in a route by which lead from Illinois could be brought to Boston entirely by water. And with Boston a day closer to Europe, they believed they might fairly challenge the supremacy of the port of New York.[69]

De Witt Clinton's counterpart in Massachusetts was Governor Levi Lincoln. Lincoln not only endorsed the report of the commissioners, but in 1826 he announced that canals were superior to railroads just as the prospect of railroad building appeared on the horizon along the Mohawk in New York, in Pennsylvania, and in Maryland. Moreover, in New York the Champlain Canal was already claiming the major share of the trade of western Massachusetts and Vermont, sending it down the Hudson to New York. Indeed, for its tonnage and its outlet for the products of New England, it was really New England's greatest canal.[70] But the rise of Troy on the eastern shore of the Hudson seemed evidence enough that Erie Canal commerce would cross that river and follow a Massachusetts canal to Boston and the Atlantic.

In 1829, however, just as plans for a canal from Boston to the Hudson were being enthusiastically discussed, the challenge of railroad interests in Boston prevented the state from such an ambitious, or foolhardy, undertaking. Instead, the Western Railroad Company was chartered in 1833. It did not receive state aid until 1836, but six years later, in 1842, through rail connections between Boston and Albany were available. The railroad had scotched plans for a canal already doomed by geography. Moreover, an improvement of Boston's ocean trade with New York and the South Atlantic seemed impossible. The long-envisioned Cape Cod Canal was surveyed in 1824 but would not be constructed until after the Civil War.

The delay in attempting a canal from Boston to the Hudson saved Massachusetts from the sobering reality of a cross-mountain canal like that built by Pennsylvania. Stimulated by the example and the rivalry of the Erie Canal fully as much as were New Englanders, the Pennsylvanians plunged ahead to surmount their mountains to the west by building canals rather than await the uncertain development of the railroad. A vast canal enterprise would bring exhilarating triumphs, solid returns, and near bankruptcy to the state.

3

Mid-Atlantic Network: Pennsylvania and New Jersey

Rivalry with the Erie Canal became most acute in Pennsylvania and the mid-Atlantic states. The resulting network of canals joined the Ohio River and the bays of the eastern seaboard, linked the Pennsylvania and Maryland coal fields with eastern harbors, and carried products to the door of New York City. Running from Pennsylvania east to New Jersey and southeast to Maryland and emerging from mountainous terrain, the mid-Atlantic canals became the most concentrated regional canal network in the nation.

Pennsylvania, with borders stretching from the Ohio River to the Atlantic, responded to the Erie Canal with panic and desperation. Her challenge was to get a transportation route to Pittsburgh on the Ohio, draw trade from Ohio and northern Pennsylvania away from the Erie Canal and the Hudson, and reverse the direction of Ohio River trade from its long-established outlet at New Orleans. By coincidence, it was only in 1817, the year the Erie Canal was begun, that upriver steamboats on the Ohio reached Pittsburgh in significant numbers. These numbers rose when the canal completed in 1831 easily enabled them to pass the falls of the Ohio at Louisville.[1]

The Chesapeake region to the south confronted Pennsylvania with still other threats in competition for western trade. Baltimore merchants built a railroad along the Potomac to Cumberland and west to the Ohio River at Wheeling. Moreover, the long-standing project of the Chesapeake and Delaware Canal would create a new waterway that would have its western terminus only a few miles south of Philadelphia. And Virginia mounted its challenge for the same western trade with the Chesapeake and Ohio Canal. Altogether the threats to Pennsylvania from the South were almost as real as that from New Yorkers to the north.

Pennsylvania had been one of the first states to build canals in hope of improving its water communications well before the challenge from New York and the Chesapeake waterways arose. It had begun its pioneer canals to bring greater regional unity and to draw its western mountainous sections into closer economic contact with the Philadelphia hinterland. The first step for early Philadelphia planners was to create a link westward from the Schuylkill Valley to the Susquehanna Valley. Beyond this they sought improved water navigation that would extend not only to the upper branches of the Susquehanna and Allegheny valleys but reach the Ohio River and Lake Erie as well. They wanted to capture the grain trade on the Susquehanna, and the discovery of anthracite coal in eastern Pennsylvania in 1791 made transportation development imperative. In the latter decades of the eighteenth century talk of such improvements had obsessed the leaders of the Pennsylvania Society for the Promotion of Roads and Inland Navigation, Governor Thomas Mifflin, the scientist David Rittenhouse, the financier Robert Morris, and the provost of the Academy of Pennsylvania, William Smith.[2]

After the failure of the pioneer efforts directed by William Weston in the 1790s, the project for a canal between the Schuylkill and Susquehanna valleys took new life when the Union Canal Company was chartered in 1811. It was to unite these two great river valleys, sometimes referred to as the Golden Link, but more specifically it was a union of the older Schuylkill and Susquehanna Company and the Delaware and Schuylkill Navigation Company.

The War of 1812 delayed the construction of the Union Canal until 1821. It was then built at the same time as the Erie Canal in New York, and its seventy-seven miles were finished in 1826. The Union Canal began near Reading on the Schuylkill River, ran up alongside Tulpehocken Creek, crossed a summit level to Quitaphilla Creek, and followed that creek and the Swatara River to the Susquehanna at Middletown. It cost $6 million, almost as much as the 363-mile Erie Canal, completed just a year earlier.[3]

The first engineer on the Union Canal was Loammi Baldwin, Jr., who predicted that it would "do for Pennsylvania and Philadelphia what the Middlesex Canal did for Massachusetts and Boston."[4] Baldwin was valuable because of his New England engineering experience, but his individualistic style brought him into conflict with William Strickland, who was a member of the company and favored the English model for smaller locks, accommodating smaller boats and lighter loads. The

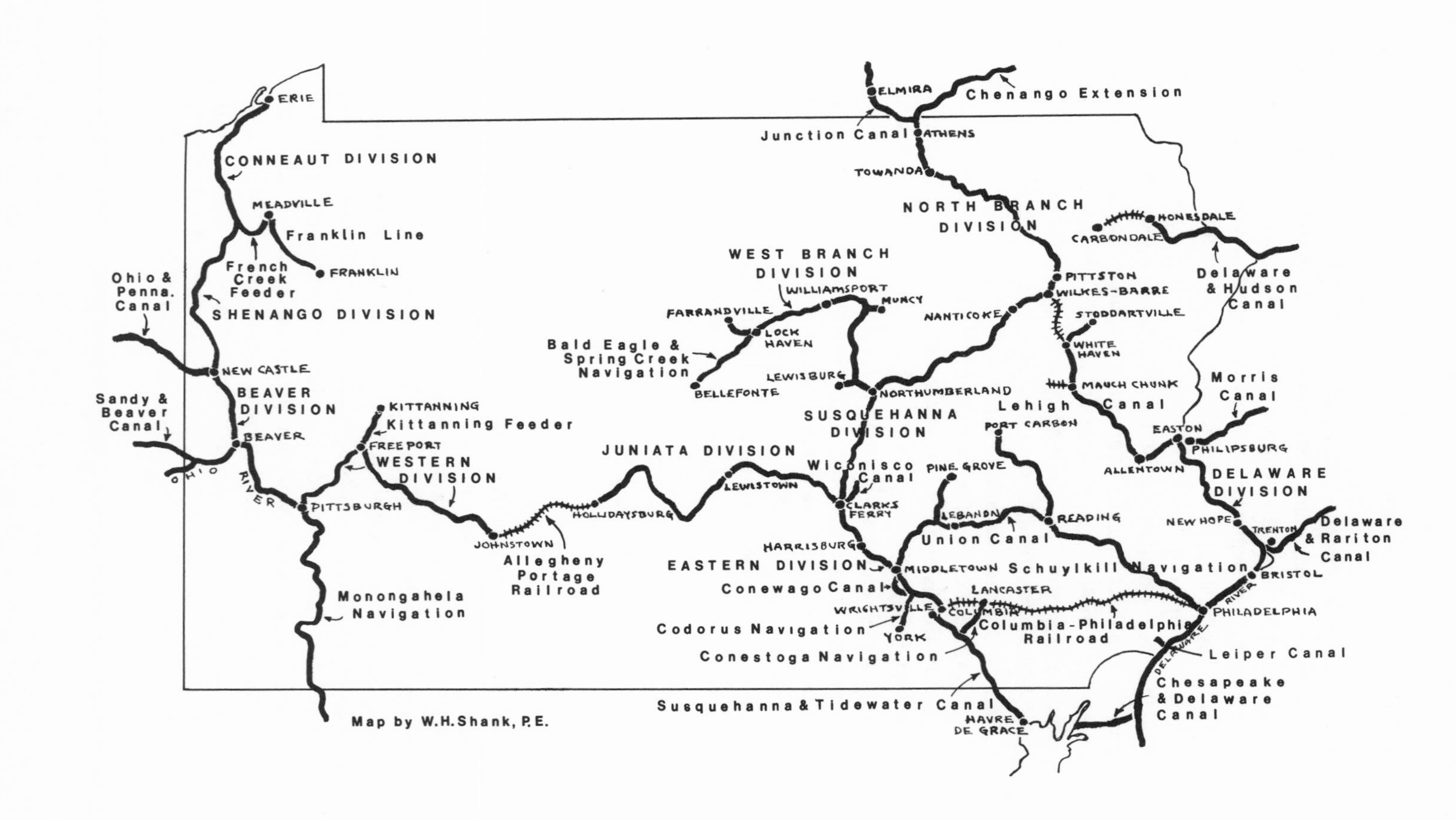
ERIE
CONNEAUT DIVISION
MEADVILLE
Franklin Line
FRANKLIN
French Creek Feeder
Ohio & Penna. Canal
SHENANGO DIVISION
NEW CASTLE
BEAVER DIVISION
Sandy & Beaver Canal
BEAVER
OHIO
RIVER
PITTSBURGH
KITTANNING
Kittanning Feeder
FREEPORT
WESTERN DIVISION
Monongahela Navigation
JOHNSTOWN
Allegheny Portage Railroad
HOLLIDAYSBURG
JUNIATA DIVISION
LEWISTOWN
Bald Eagle & Spring Creek Navigation
BELLEFONTE
FARRANDVILLE
LOCK HAVEN
WEST BRANCH DIVISION
WILLIAMSPORT
MUNCY
LEWISBURG
NORTHUMBERLAND
SUSQUEHANNA DIVISION
Wiconisco Canal
CLARKS FERRY
HARRISBURG
EASTERN DIVISION
MIDDLETOWN
Conewago Canal
WRIGHTSVILLE
COLUMBIA
YORK
Codorus Navigation
Conestoga Navigation
LANCASTER
Columbia-Philadelphia Railroad
Susquehanna & Tidewater Canal
HAVRE DE GRACE
PINE GROVE
LEBANON
Union Canal
READING
Schuylkill Navigation
PORT CARBON
Elmira
ELMIRA
Chenango Extension
Junction Canal
ATHENS
TOWANDA
NORTH BRANCH DIVISION
PITTSTON
WILKES-BARRE
NANTICOKE
STODDARTVILLE
WHITE HAVEN
MAUCH CHUNK
Lehigh Canal
CARBONDALE
HONESDALE
Delaware & Hudson Canal
Morris Canal
EASTON
PHILIPSBURG
ALLENTOWN
DELAWARE DIVISION
NEW HOPE
TRENTON
Delaware & Rariton Canal
BRISTOL
RIVER
PHILADELPHIA
DELAWARE
Leiper Canal
Chesapeake & Delaware Canal
Map by W.H.Shank, P.E.

Massachusetts engineer was replaced in 1823 by the New York engineer Canvass White, who brought with him three of his Erie Canal assistants and who could more readily work within the company organization.[5]

The conflict between Baldwin and the company cast a long shadow over the future of the Golden Link between the Schuylkill and Susquehanna valleys. At the six-mile summit level near Lebanon, the porous limestone rock required a constant infusion of water. Baldwin would have made a deeper and lower canal, with larger locks, costing more than the company was willing to bear. White found an easier route, less costly in the short run. But the steadily leaking summit of White's design required feeder after feeder, three reservoirs, and steam engines at the pumping station near Jonestown. There the waterworks lifted water a hundred feet above the canal, where it dropped through wooden pipes for three miles into the canal. Ultimately, the entire summit-level channel was lined with plank.[6]

The Union Canal had ninety-four locks and two aqueducts, but its most dramatic feature was the 729-foot tunnel two miles west of Lebanon. It was 18 feet wide and 16 feet high, pierced through solid rock, and was the second canal tunnel in the nation. It would be widened and heightened in the 1850s when the Union Canal was enlarged.

The last boat would not pass through the Union Canal tunnel until 1885, yet the small locks favored by William Strickland, measuring seventy-five feet by eight and a half feet, and its thirty-six-foot channel left this first counterpart of the Erie Canal fatally flawed. One of its flaws was that it was too small to take the boats that used the Schuylkill Navigation system, a 106-mile slackwater and canal navigation on the Schuylkill River from Philadelphia to the coal fields at Port Carbon, which took boats twice as large as those that could be used on the Union Canal.[7]

Because of the small dimensions of the Union Canal, when the Mainline Canal was built west to the Ohio River, the new system would virtually bypass the older but still operating Union Canal. In the Mainline system, combining canals and railroads, the Columbia and Philadelphia Railroad was built almost due west to the Susquehanna, well below the Union Canal. And even this route would be shunned when a new canal was built along the lower Susquehanna in 1840, the Susquehanna and Tidewater Canal, which began at Wrightsville opposite Columbia (the railroad terminus) and went along the river to Havre de Grace on the Chesapeake above Baltimore. From there, boats could cross the head of the bay to the Chesapeake and Delaware Canal after 1830, making a short connection to the Delaware River, on which boats could pass to Philadelphia.[8]

The Union Canal, however, was to a large extent the completion of an earlier project that had been delayed by lack of funds and the War of 1812. Pennsylvania entered the Canal Era most ambitiously with the Pennsylvania Mainline Canal in 1826, the year the Union Canal was completed. The Pennsylvania Mainline Canal would go all the way to the Ohio River at Pittsburgh, far to the west of the Schuylkill Navigation and the Union Canal. To begin such an undertaking was a decision of great consequence for Pennsylvania, with results beyond description.

With the Pennsylvania Mainline Canal Pennsylvania embarked on a project that has been called "that technological monstrosity, a canal over the mountains."[9] It was to be the trunk line of the most extensive canal system in the East, facing awesome challenges, bringing dramatic achievements, pursued with audacity, and accomplished with almost profligate expenditure. The Mainline would follow the Susquehanna and Juniata rivers to the mountains, cross them via a portage railroad, and then follow the Allegheny and Conemaugh rivers to Pittsburgh. Added to the Mainline would be a network of anthracite canals to carry coal to the seaboard and provide linkage to the waterways of New York.

In retrospect, this audacious plan of canal building in such mountainous country has been judged a reckless failure. Yet Pennsylvanians believed they were taking a conservative, imitative course, following the successful example of New York and the Erie Canal, and they were forsaking the more innovative option just then appearing in the new railroad technology.[10] It was the new railroad technology, still unproven in its capacity to surmount steep grades or carry heavy freight, that was gambled on by the Baltimore investors who have been so often praised for inaugurating a railroad route to the Ohio River with the Baltimore and Ohio Railroad.

The idea for a canal from Philadelphia to Pittsburgh, of course, had its roots in the plans of the Society for the Promotion of Roads and Inland Navigation. In 1824 Mathew Carey set out to create a similar organization in the Pennsylvania Society for the Promotion of Internal Improvements, which became the primary force in moving the legislature to authorize an east-west canal two years later. Carey enlisted the support of John Sergeant and some fifty-five others by 1826 to work for a canal linking the Susquehanna and the Ohio. Their great fear that Philadelphia might lose out in the commercial rivalry from New York and Baltimore fueled their efforts as they managed the canal project when it went before the legislature for action.

If the society became in reality a canal lobby working through its Acting Committee and a vehicle for Carey's program for economic development, its announced purpose was to gather technological information on canals and railroads. The society sent William Strickland to England to study canals and railroads, much as Canvass White had been sent by New Yorkers a decade earlier. Although Strickland gave special attention to English railroads, Carey and the society had a clear bias toward canals. They used the information they gathered selectively to serve their goal of a great western canal, which they believed essential for the commercial growth of Philadelphia.[11]

Carey and the society poured forth material on canals which fed an almost continuous public debate. The society published a series of pamphlets, most of which called for a canal but carefully avoided designating a particular route. They predicted rich returns and showed how the canal might be financed. But on one theme they were unequivocal: the waterway must be constructed by the state. No private company could undertake such a task.

Carey added innumerable pamphlets of his own, which gained force from his long career as a publicist and advocate of a nationalistic program such as the American System of Henry Clay. His forensic skill had already helped create the Chesapeake and Delaware Canal.[12] Writing under his own name, he could be less restrained in his advocacy of a canal than were the pamphlets published by the society. A flood of arguments for an east-west canal appeared in pamphlets, petitions, and newspapers and circulated in thousands of copies.

Mass meetings were held throughout the state. Carey arranged a great town meeting in Philadelphia in January 1825 to publicize his canal ideas. The meeting resulted in a Philadelphia memorial for "complete water communication" from the Susquehanna to the Allegheny rivers to be built by the state. It bore a thousand signatures and was almost identical to the one written by De Witt Clinton for the Erie Canal in 1816.[13]

Carey and the society feared the divisive response to a debate on the crucial question of whether to build a railroad or a canal, and they feared that a railroad would prove more beneficial to Baltimore than to Philadelphia. But John Stevens had secured a charter in 1823 for a railroad from Philadelphia to Columbia on the Susquehanna, and he now pressed for a longer line. Each new bit of information on the success of English railroads between 1823 and 1826 was published in the press. Pamphlets favoring a railroad made it clear that the options were still open, and they contained ample warnings that Pennsylvania could not hope to reap the same benefits from a canal to the West as those won by New York, no matter how great the zeal of canal advocates. But the latter were able to keep serious consideration of a railroad at bay until sufficient support for a western canal could be won.[14]

This essential support for a canal was gained at the Harrisburg Internal Improvement Convention in August 1825. William Lehman, John Sergeant, and others won support for resolutions favoring the canal plan, with railroads only as auxiliary connections, and calling for immediate construction of the Mainline Canal. Although the exact route remained unsettled, the cost was now estimated at $8 million, and eight years might be required for construction.

But the convention also heard the strident voices of sectional opponents, much as sectional divisions in New York had opposed the decision to build the Erie Canal. Opposing delegates from six southern counties represented interests tied to Baltimore, while delegates from northeastern counties tied to New York also cast their votes in the negative. But the canal forces had the majority, and their victory in the convention presaged their victory in the legislature.

Just when the Harrisburg convention had backed a canal, however, William Strickland's report on his examination of canals and railroads in England was made public. His conclusion was that "railroads offer greater facilities for the conveyance of goods, with more safety, speed and *economy.*"[15] But when pressed for details, Strickland equivocated on the use of railroads in the mountains, recommending stationary engines and inclined planes instead. Carey now publicized his opposition to railroads, emphasizing the lack of experience with locomotives on grades and the high cost of railroads.

Such debates were ultimately aimed at influencing the legislature. Lobbied by the society for the Promotion of Internal Improvements, and led by William Lehman, the influential Philadelphian who had long worked for an east-west canal and now chaired the Inland Navigation Committee in the House, the legislature appointed a board of commissioners for internal improvements in 1824. John Sergeant was the president of the new board, and its report in 1825 argued "the perfect practicality of making a canal" on the route from Philadelphia to Pittsburgh. At last a canal route was projected: from Philadelphia to the Susquehanna above Harrisburg, then to the Frankstown branch of the Juniata near Hollidaysburg, through a four-mile tunnel to the Little Conemaugh River, and on to Pittsburgh. A second legislative act in 1825 appointed a Board of Canal Commissioners.

William Lehman was the De Witt Clinton of Pennsylvania. He seized upon the momentum of the Harrisburg convention of 1825, as well as information publicized by the society and by Carey, to steer the canal bill toward enactment. A crucial compromise was made when canal advocates accepted a railroad portage over the mountains and the canal commissioners recommended construction only on the nonmountainous sections along the Susquehanna and Allegheny rivers. Far from working for the railroad party, as might have been expected, William Strickland was hired as an engineer for the canal, and he prepared the maps and estimates for the routes under consideration.

Early in 1826 the legislature moved to begin construction. The Board of Canal Commissioners made its report, a House resolution called for a decision, and a bill was introduced in the state senate "to provide for the commencement of a Canal" at the expense of the state. The waterway was "to be styled the 'Pennsylvania Canal.'"[16] Six weeks later the key elements of the Pennsylvania Mainline Canal were enacted into law.

Unlike New York, which began construction of the Erie Canal in the middle, Pennsylvania authorized construction only at the extremities of the canal. In the east this was from the terminus of the Union Canal along the Susquehanna to the Juniata River and in the west from the mouth of the Kiskiminetas River to Pittsburgh, amounting altogether to only forty-four miles of canal. Yet the canal law of 1825 also provided for an extension of the canal from Pittsburgh over the high ground to Lake Erie, by whatever navigable route the commissioners might find.

Work began on July 4, 1826, near the capitol at Harrisburg on the Eastern Division, ironically the division that would be the last to be finished in 1833. Here the line between the Swatara River at Columbia and the Juniata River was located by William Strickland, whose identification with the canal went back to his work with the Pennsylvania Society for the Promotion of Internal Improvements. The Erie Canal engineer Nathan S. Roberts was brought to locate the canal line on the Western Division between Pittsburgh and Freeport at the mouth of the

Kiskiminetas River. Other New York engineers soon followed, notably James Geddes and Canvass White, bringing the expertise gained on the Erie Canal to build Pennsylvania's rival waterway. The Mainline Canal was modeled on the Erie, forty-five feet wide and four feet deep, with locks ninety feet long and fifteen or seventeen feet wide.

In the second canal law in 1827 the Pennsylvania legislature authorized two more segments of the Mainline route for construction: the Juniata Division east along the Juniata River from its mouth on the Susquehanna to Lewistown and the Western Division from Blairsville to Freeport along the Kiskiminetas and Conemaugh rivers. These links along the Juniata and the Kiskiminetas began to fill in the canal, but the law of 1827 also made a radical departure from the primary promotion of the great east-west canal envisioned in the origins of the Pennsylvania Canal.

Under the new law, the commissioners were to begin work on the feeder from French Creek to Conneaut Lake, begin surveys on canals between Pittsburgh and Lake Erie, continue a canal on the Susquehanna Division up the Susquehanna past the mouth of the Juniata to Northumberland, and begin a canal on the Delaware Division along the Delaware River between Bristol and Easton. As the first historian of the Pennsylvania canals, Avard L. Bishop, wrote in 1907, this was "the commencement of a complete change of policy," foreshadowing the great expansion of the Pennsylvania canals.[17] Wholesale canal building, often pursued more extensively off the Mainline than on that primary trunk route, allowed only incremental progress on Pennsylvania's great canal to the West. Like the delay at the mountain ridge at Lockport on the Erie Canal, a solution to crossing the Alleghenies by a portage railroad west of Hollidaysburg would not be decided upon until five years after the Mainline was begun. The precedent for this portage by railroad was found in the use of an initial railroad segment that would start the Mainline route in the east.

The Mainline was originally designed to link up with the Union Canal at the Susquehanna River, which would connect to the Schuylkill Navigation and Philadelphia. But that canal was found to be too small, and its limestone summit had an inadequate water supply. In its place an eighty-two-mile horse-powered railroad was begun in 1828 by the state, and after 1834 the Mainline Canal began not at the Union Canal terminus but farther down the Susquehanna River at the new railroad terminus at Columbia. A full description of this pioneer railroad might be minimized in a canal history, but it was operated as part of a canal and served as an integral part of the Mainline system. An early lithograph shows a railroad car with a section of a passenger canal boat loaded on it, departing from Third and Dock streets in Philadelphia, ready to pass up Broad Street to the Columbia and Philadelphia Railroad and on to the Mainline Canal.[18] Baldwin steam locomotives were in use when the double-tracked road was completed in 1834, but for another decade they shared the tracks with horse-drawn vehicles.

In the beginning, when the power of locomotives was limited, two inclined planes using stationary steam engines were included on the route. One hoisted the railroad cars up from the Schuylkill River to Belmont, and to the west, another dropped from Columbia to the bank of the Susquehanna at the end of the railroad. They were not replaced by locomotives of sufficient power until 1836 and often caused long delays. In the meantime, cables attached to the cars operated on the double-tracked line by a pulley so that the descending car pulled up an ascending car on a parallel track. Similar inclined planes would be the backbone of the portage railroad at the crest of the Alleghenies when that segment of the Mainline was built.

As Albright Zimmerman has observed, this short railroad is often overlooked because it was built and operated by the state and run by canal commissioners. It was the first double-track railroad in the nation, and it operated more steam engines (thirty-four in 1836) than any other American railroad at the time. East of Lancaster it passed over a 60-foot-high viaduct 804 feet long, and west of Lancaster it crossed a seven-span railroad bridge over the Schuylkill River which was more than a thousand feet long. Although it was only the first segment in the trunk of the Mainline, it carried a great deal of traffic. In 1835, 1,118 engine trips pulled 10,588 cars at a time when it still shared the road with horse-drawn vehicles as a public highway. In 1847 the Pennsylvania canal commissioners calculated its commerce at 234,229 tons.[19]

At the end of the Columbia and Philadelphia Railroad on the Susquehanna River construction lagged on the Western Division up the Susquehanna to the Juniata River as funds were diverted to other canals. And at Hollidaysburg in the high mountains, the route of the capstone link of the Allegheny Portage Railroad had still to be determined. The most intensive canal-building was therefore on the Western Division, where work began at Pittsburgh in 1827 and moved eastward along the Monongahela and Allegheny rivers, up to the Kiskiminetas, and up that stream to the Conemaugh to Johnstown. The western end of the Allegheny Portage Railroad when it was finally completed was at Johnstown.

Construction on this western extremity of the Mainline was as dramatic as any canal building in Pennsylvania thus far. The 104-mile canal line here, which was located by the New York engineer Nathan S. Roberts, began at the Monongahela River and ran eastward to Johnstown. It required a tunnel, sixty locks, and sixteen aqueducts, almost as many structures as Roberts had seen on the entire length of the Erie Canal.

Yet these figures do not do justice to the achievement of constructing a canal in difficult terrain, when party politics produced a succession of three superintending engineers. In 1827 an economizing legislature lowered the engineers' salaries from $3,000 to $2,000 a year, which led to the resignation of Nathan Roberts, as well as William Strickland, James Geddes, and David B. Douglass elsewhere on the Mainline. Nathan S. Roberts was succeeded on the Western Division by his young assistant James D. Harris, whom Roberts had trained on the job. Harris fell afoul of the Jacksonian politics of the commissioner in charge of the division, and he was replaced by Sylvester Welch. The latter had come with Canvass White from New York to the Union Canal, and he carried the Western Division through to completion.[20]

An account of the political impediments to the progress of construction has been left by Solomon W. Roberts, an eighteen-year-old engineer working on the section of the Western Division between Blairsville and Johnstown for $2 a day. Roberts had previously worked for his uncle Josiah White with Sylvester Welch on the Lehigh Canal. In 1829 he began a six-year service on the Pennsylvania state system. In his "Reminiscences" written almost fifty years later, he remarked on the way politics hindered construction. "The Canal Commissioners were politicians," he wrote, and in the competition for contracts canal work "often failed to endure the strains to which it was subjected." He was "far from being pleased with the general results of my experience on the canal in the valley of the Conemaugh."[21]

The Pennsylvania canal law of 1826 designated Pittsburgh on the Ohio River as the terminus of the Mainline. But the implementation of this simple statement was as complex as that of the termini of most canals of the Canal Era. Surveys put the canal on the west side of the Allegheny River, making it necessary to bring the canal into Pittsburgh over an 1,140-foot aqueduct resting on seven piers, which was completed in

1829. Imposing as this structure appeared, fears that it might collapse after several years of use resulted in its replacement by John A. Roebling's famous suspension bridge aqueduct, the first such structure in America.[22]

And in Pittsburgh a great tunnel had to be dug before a terminus harbor could be reached on the Monongahela River. The canal entered Pittsburgh through the aqueduct crossing the Allegheny onto the triangular point between the Allegheny and Monongahela rivers before they joined to form the Ohio River. To gain access to the Monogahela (and to connect with the Chesapeake and Ohio Canal when that waterway was finished), the canal was dug under Grant's Hill, through an 810-foot tunnel which ran deep underground and dropped by four locks to a terminus in the Monongahela. When the Grant's Hill tunnel was finished in 1831, it ranked with that on the Union Canal, completed only a few years earlier, as one of the first canal tunnels in the nation.[23]

Soon the facilities needed to serve the canal terminus began to appear in Pittsburgh. Canal basins, a weighlock, houses for toll collectors and lock tenders, warehouses for freight, and more were built. Steamboats and river traffic still dominated, but Pittsburgh became a canal port as well. The sight of the city bisected by the canal, the Allegheny aqueduct, the canal tunnel, and the ascending locks might well have dazzled travelers approaching the Mainline from the West. New packet lines appeared on the Western Division and brought passengers who entered the great Ohio Valley from the East. David Leech began the Western Transportation Company in 1830 and pioneered packet travel west of Johnstown.

If the majestic aqueducts were the glories of the Erie Canal in New York, the tunnels on the Western Division of the Mainline were its most imposing features. Between Pittsburgh and Johnstown the Mainline Canal passed by the towns of Freeport, Leechburg, and Saltsburg on the Kiskiminetas River and Blairsville, Lockport, and Fairfield on the Conemaugh River. At Tunnelton below Blairsville on the Conemaugh, a thousand-foot tunnel opened directly into the aqueduct across the Conemaugh. Philip Holbrook Nicklin traveled the canal eastward in 1835 and described this astonishing joining of aqueduct, tunnel, and mountainous canal:

> We passed over a beautiful stone aqueduct which leads into the mouth of a large tunnel eight hundred feet long which perforates the mountain and cuts

> off a circuit of four miles. The tunnel is cut through limestone rock for four hundred feet, and the rest is arched with solid masonry, as are both the entrances. . . . You are gliding over the aqueduct . . . and then to your great astonishment you perceive an enormous archway which passes through the mountain's base, and discover the brilliant landscape beyond, set in a dark frame composed of massy ribs of rock dimly seen within the tunnel, upon which the mountain rests. Directly after leaving the tunnel the boat enters a pool made by building a dam across the river, . . . passing through the woods and air and water in the sweetest silence, save the lulling sounds of tiny ripples at the bow.[24]

On the Conemaugh itself the Johnstown Dam was constructed to supply water for that part of the canal. This dam would be swept away later in the great Johnstown flood. Despite all the difficulty in constructing these structures, the Western Division was opened for navigation in 1831, three years before the high mountain section above Johnstown and Hollidaysburg had been begun.

The high mountain section rose between Hollidaysburg on the eastern slope of the Alleghenies and Johnstown on the western slope. There the Allegheny Portage Railroad surmounted the mountains on a thirty-eight-mile route which had been left undecided as construction moved up the mountains from both east and west. The technology of the stationary steam engine already in use on the planes of the Columbia and Philadelphia Railroad at the eastern end of the Mainline made it possible to pull cars carrying section boats for the climb over the mountains. After surveys by Nathan S. Roberts and other engineers, construction was supervised by Moncure Robinson, Sylvester Welch, Solomon W. Roberts, Edward Miller, and W. Milnor Roberts.

Moncure Robinson was the engineer who provided the basic design for the Portage Railroad. A Virginia engineer, he had worked on the James River and Kanawha Canal from 1823 to 1825, and he had spent the summer of 1822 studying firsthand the construction of the Erie Canal. In 1825 he went to Europe to see its canals, especially those in France and England. But it was English railroads that excited him, and he returned in 1827 advocating railroads in the United States. The Pennsylvania canal commissioners gave him the challenge of designing the Allegheny Portage Railroad. He worked with army engineers Colonel Stephen H. Long and Major John Wilson and provided a plan for a railroad, a tunnel, and inclined planes. It was Robinson's plan, without the tunnel, that Sylvester Welch followed when Welch was put in charge of constructing the Portage Railroad.[25]

Often deprecated by critical accounts emphasizing the failure of the Pennsylvania canal system, the Allegheny Portage Railroad was one of the great engineering triumphs of the Canal Era. The achievements in clearing, grading, cutting, and embankment and the construction of roadbed and track, often on sheer precipices, were amazing triumphs of innovation and invention. From Hollidaysburg west, the railroad rose almost 1,400 feet in ten miles to its highest point 2,334 feet above sea level, and on the western slope it dropped 1,171 feet in twenty-six miles to Johnstown, where the Western Division began.

There were ten inclined planes, five on each slope of the mountains, and tracks were placed for horse-drawn cars between the planes. On the ten inclines, parallel tracks carried upward and downward cars, pulled by an endless rope eight inches around. When possible, cars being pulled up the incline were attached with cars descending the incline to balance the weight as was being done on the Columbia and Philadelphia Railroad. Three or four boat-carrying cars could be pulled at once in each direction, moving on inclines where the pitch varied from 6 to 10 percent.

The levels and grades between the inclines were double-tracked after 1835, and in the beginning the state operated the inclines only. Between them the road was a public highway with horse-drawn cars. Later the state furnished locomotives and cars as well. English railroad technology was drawn upon again when Edward Miller was sent there in 1831 and then returned to supervise the operation of the stationary engines. John A. Roebling's inventive genius worked again when the hemp rope was replaced with his great wire-woven cables, the same technology he applied to his suspension bridge aqueducts.[26]

Four miles east of Johnstown, a 900-foot tunnel took the railroad under the mountains. It was the first railroad tunnel in America, completed in 1834, and was arched and fitted with stone at each end. Its construction was supervised by Solomon W. Roberts, whose reminiscences describe the conditions overcome by the builders of the portage railroad. Now a young man of twenty, Roberts was an assistant engineer working for Sylvester Welch, beginning in April 1831. As his party was finding a line for the railroad near the summit, the mountain wind blew so strongly that it took bark from dead trees. They slipped in snow and slept in tents under buffalo robes. When summer came, gnats were so troublesome that the men worked in the smoke of burning leaves and greased their faces. Later as many as two thousand men labored at a time on the portage railroad line, clearing a 120-foot swath through stands of trees 100 feet high. Axmen were so troubled by rattlesnakes that they caught some live to impress Governor George Wolf when he visited their camp.[27]

Solomon Roberts was given supervision of the work on the western slope. (W. Milnor Roberts was in charge of the eastern part of the portage.) Solomon Roberts designed the great horseshoe bend, or Conemaugh viaduct, a semicircular arch seventy feet high with an eight-foot span. But his reminiscences record that the Scottish stonemason who helped build it, John Durno, fell to his death as he worked on another high bridge. Roberts also helped design and anchor the planes, he managed the importation of rails from England, and in 1835 he was put in charge of the operation of the entire portage route.

From the windows of his office at Ebensburg on the turnpike near the line, Roberts watched as "the Conestoga wagons loaded with emigrants, with their baggage and furniture, slowly wended their way to the West." His portage railroad would eliminate this slow and difficult journey. In his reminiscences he wrote, "We were striving to build a great public work to endure for generations, and as it turned out, it was superseded by something better in about twenty years."[28] In 1854 the mountain division of the Pennsylvania Railroad was opened, with a tunnel under the summit, and the portage railroad with its remarkable inclined planes was abandoned.

Roberts recorded chance encounters in 1835 when he dined at the summit with the new governor-elect, Joseph Ritner, on his way to Harrisburg, and with Henry Clay and Felix Grundy, who were going to Washington for the opening of Congress. The list of the famous and influential who used the Mainline with its portage railroad is long. Ulysses S. Grant took the Mainline on his way to enter West Point in 1839. Political leaders such as Thaddeus Stevens, James K. Moorhead, and John W. Geary took the route, as did the famous singer Jenny Lind and a host of foreign travelers.

Charles Dickens, on a trip west in 1842, described the rails of the Portage Railroad "laid upon the extreme verge of a giddy precipice" as he looked down on tiny figures "and we riding onward, high above them, like a whirlwind."[29] Traveling eastward in 1835, Philip Holbrook Nicklin reached the summit of the Allegheny Portage Railroad and described "the fear of the steep descent which lies before you and as the car rolls along on this giddy height the thought trembled in your mind that it may slip over the head of the first descending plane, rush down the frightful steep and be bashed into a thousand pieces."[30]

As if the steep slopes of the Portage Railroad were not startling enough to the observer, the sight of sectional canal boats being hauled over the mountains was common in the 1840s. These were boats divided into four sections, mounted on cradles set on double-wheeled trucks. Separated into sections on the planes and intervening grades, they were joined together to form a single boat on the canals. When they were used on the Columbia and Philadelphia Railroad at the eastern end of the Mainline, they rode on their railway trucks, even on the streets of Philadelphia, where they gathered their cargoes of manufactures and merchandise.

Commonplace as they became, the section boats evolved slowly and were brought into use only after conflict and delay involving bitter personal contests and litigation. They were developed chiefly by John Dougherty in response to the peculiar problems of the Mainline Canal, which required five transfers of freight between Philadelphia and Pittsburgh. Dougherty was a former canal contractor and partner in the Reliance Transportation Company at Hollidaysburg, and he adapted the ideas of John Elgar of Baltimore for sending watertight tanks over the canal and perfected the section boat, which could be hauled on cars. Dougherty promoted his section boats in his own newspaper, proposed that locomotives be individually owned to prevent monopolies, and wrote that his system would tend "to allay the violence of party spirit."[31]

Before his innovation could be standardized into a four-section boat and the state would agree to pull the boats on the inclines, Dougherty engaged in bitter conflict with his builder and then with the canal commissioners over their use. The commissioners feared their effects on Mainline traffic because they could carry only six to ten tons and were narrow so they could pass each other on the Allegheny Portage Railroad. The state insisted on using its own trucks rather than renting those developed by Daugherty. Section boats were used chiefly on the Mainline, and eight hundred passed east and west on the Portage Railroad in 1849 alone. But they failed to solve the transfer problems of the Mainline Canal. As Jesse L. Hartman concluded, "In the economics of canal-railroad transportation the section boat did not find a lasting place."[32]

When the Mainline route was completed in 1835, nine years after it had been started, it included the successive segments going west of the Columbia and Philadelphia Railroad, the Eastern and Juniata canal divisions, the Allegheny Portage Railroad, and the Western Division. Altogether the route stretched 395 miles, had 174 locks and forty-nine aqueducts, and passed through three tunnels. Its original cost was a staggering $12,106,788.[33] The high initial cost and later financial failure, however, have obscured the considerable success of the Mainline Canal.

Traffic was heavy in the 1830s and 1840s and, on some parts of the canal, through the 1850s. The Allegheny Portage Railroad, for example, carried 29,740 tons west and 15,439 tons east in 1836 and almost 85,000 tons each way in 1846.[34] Cargoes were much the same as those on the Erie Canal: wheat, flour, salt, whiskey, and lumber going east and manufactured goods taken west. Records of a country store kept by Peter O'Hagan in Newry near Hollidaysburg from 1842 to 1847 showed that he was supplied from the western section of the Mainline with whiskey, glass, molasses, and sugar, much of it purchased from a wholesale grocer in Pittsburgh.[35] On the Western division the land rate of eighteen cents a mile for freight dropped to three cents a mile in 1834. That year, the *Philadelphia Price Current* quoted the cost of bringing goods from Cincinnati to Philadelphia at $1.30 per hundred pounds and the cost from Cincinnati to Albany on the Erie Canal at $2.40.[36] The time required for freight passage from Philadelphia to Pittsburgh was reduced to seven days.

The packets of the Western Transportation Company operated by David Leech were the first on the Western Division in 1830, and he introduced others such as the *William Lehman* on the Juniata and Eastern divisions in 1832. His became the most successful packet line on the Mainline Canal. His cars left Philadelphia on the Columbia and Philadelphia Railroad, his packets ran on the canal from Columbia to Hollidaysburg, his cars were pulled over the Allegheny Portage Railroad, and his packets continued west from Johnstown to Pittsburgh. He kept stables every ten to twelve miles for changing horses. The Leech transportation line ran as long as the state operated the Mainline Canal, and his offices became centers of political power in canal affairs until he died in 1858.[37]

Nine packet lines operated on the Mainline Canal in 1837, and the leading companies were Leech's Western Transportation Company, the Express Line, and the Pioneer Line. In the month of April alone, the collectors at Hollidaysburg and Johnstown reported that these three lines carried 3,091 passengers going west, for 111,233 passenger miles, and 1,313 passengers traveling east, for 47,320 passenger miles. If more freight came east on the canal, more passengers went west. The total passenger miles traveled westward from Johnstown on the canal in 1846 numbered 953,317; those coming eastward, 274,790.[38] The fare was $6 between Philadelphia and Pittsburgh, or two cents a mile. The Pioneer Line made the through trip in three and a half days in 1836. Boats left Columbia at the end of the Columbia and Philadelphia Railroad and made connections for Middletown, Harrisburg, Hollidaysburg, and the Portage Railroad, and then from Johnstown to Pittsburgh. Harrisburg became a city of canal basins. Although most immigrants went west on the Erie Canal, thousands traveled the Mainline, often stopping at hotels in the towns and cities along the way.

Hopes that the Mainline could be a strong competitor to the Erie Canal, however, were elusive. The inconvenience of changing from railroad to canal to railroad to canal, the cost of canal offices at five places on the route, and the problems of carrying loaded boats on railway cars all discouraged through traffic. Some merchants in Philadelphia shipped to New York and then used the Erie Canal to reach western markets.[39]

The Mainline was only one part of a longer Pennsylvania system of 435 miles of lateral canals, many begun while the Mainline was still under construction. In 1831 only 50 of 155 miles of canal construction were on the Mainline, and when the Mainline was completed, an additional 243 miles of public works were also ready for use, making 637 miles in all. By 1842 an additional 135 miles of public works were finished and 162 more were started, raising the total to 934 miles.

But the cost of canal construction boosted the state debt to $32 million in 1839, and by 1843 the state debt for its public works was almost $40 million, a figure that remained almost constant until the canals were sold. Tolls on all canals brought in about $1.25 million annually, which yielded about $0.5 million over operating expenses. The Mainline cost $16.5 million by 1857, and by 1860 the total costs for the Pennsylvania public works had reached more than $100 million, more than any other state spent for internal improvements during the period.[40]

Political support for this expansion was bipartisan and rested on an essential democratic belief in the virtue of public spending. In this belief, as Louis Hartz has written, such expenditures could be justified "by the capacity and the functioning of the economic system, by the size of the population, [and] by the productive powers of the community."[41] But legislative action also reflected localism, logrolling, and elements of a spoils system offered by employment on the public works. When Thaddeus Stevens wrote the bill to charter the United States Bank as a state bank, the title of the act included the phrase "to continue and extend the improvements of the state by railroads and canals." Although party conflicts were often ferocious, the public works were supported by a succession of governors between 1829 and 1845. Governor George Wolf was the counterpart of William H. Seward in New York when he wrote in 1832 that when the works then under contract were completed, the state "will have in her own right" constructed 593 miles of canals and slackwater navigation, "which in magnificence of design, solidity and neatness of execution will stand unrivalled and will place our state upon a proud eminence."[42]

This expansion, however, was financed by borrowed money, and Pennsylvania proved especially vulnerable to the pressures of the panic of 1837 because of the expiration of the Second Bank of the United States in 1836 and the subsequent failure in 1841 of Nicholas Biddle's United States Bank chartered in Pennsylvania. The failure of the Girard Bank and the Bank of Pennsylvania in 1842 added to the state's credit crisis. Pennsylvania never repudiated its debt, but it could not meet five interest payments due between 1842 and 1844. British investors, who held two-thirds of the $40 million Pennsylvania debt, protested against the actions of the Pennsylvania legislature and attacked the character of American democracy. Sydney Smith, the British essayist, lectured Congress that it should be "teaching to the United States the deep disgrace they have incurred to the whole world."[43] In 1844 the Pennsylvania legislature responded to the growing crisis by taxing property and salaries, similar to the 1842 stop and tax law in New York. Interest payments on the public debt were resumed, but a movement to sell the public works began in the early 1840s. The sale was forestalled, but in 1857 the public works were sold at a fraction of their value.

In this sale the Pennsylvania Railroad Company purchased the Mainline Canal for $7.5 million and the canal capitulated to the railroad, which had been projected when the east-west line had been begun. Ever since it was chartered in 1846, the Pennsylvania Railroad Company had attacked the Mainline system, especially the management of the state-owned Columbia and Philadelphia Railroad. The portage railroad was abandoned immediately after the sale, though the Western Division continued in operation until 1900.

In the hindsight of history, the Mainline Canal appeared to have earned nothing for the state. Profits over operating expenditures were only $6.7 million.[44] The dire prophecy of those who had argued that Pennsylvania enjoyed no favorable route for a canal comparable to New York's had come to pass. Moreover, Baltimore stole a march to the west by railway just as New York had preceded Pennsylvania by canal. Even Boston was linked to Albany by rail before the Pennsylvania Railroad reached the Ohio.

Yet the dilemma of Mathew Carey and others remains understandable. They feared to delay until the railroad had proven itself for their mountainous state. Nor could public support for canals be ignored. Carey and others could not know that money not spent on canals would have been devoted to railroads perhaps a decade later. Moreover, it was the branch canals, the political price for continued construction on the Mainline, that appeared to bear much of the responsibility for the financial distress of the state in the 1840s.

Branch canals reached out from the Mainline to other waterways leading to the coal fields and running to the borders of the state. Calling these canals branch or lateral canals, however, is ambiguous at best. Many were connectors in the Pennsylvania canal system, but the anthracite canals were regional waterways bearing a distinct identity of their own. And the branch canals added north-south connections to the Pennsylvania system, which had begun as an east-west link to the Ohio Valley.

In the central part of the state, the Susquehanna Division followed that river north for forty miles to the canal port of Northumberland, there to divide into the North Branch Division, which connected with the New York canals, and the West Branch Division, which followed the west branch of the Susquehanna to Williamsport, Lock Haven, and Bald Eagle. Southward along the Susquehanna, from Columbia and Wrightsville, the privately built Susquehanna and Tidewater Canal made a direct connection to the Chesapeake at Havre de Grace. In the southeastern part of the state, the Delaware Division Canal ran for sixty miles along the western side of the Delaware River from Easton to Bristol at tidewater. The twelve-mile Wisconisco Canal was built from Millersburg along east side of the Susquehanna River to Clark's Ferry.

Far to the west, near the Ohio border, the Beaver Division completed the link to Lake Erie that was projected in 1825. The Beaver and Erie Canal left the Ohio River at Beaver, ran to New Castle, and continued north to Erie. The portion north of New Castle was also called the Erie Extension Canal, and connected to it was the French Creek feeder, which reached Meadville and Franklin. The French Creek feeder was authorized by the canal act of 1826 to add more water to Lake Conneaut to supply the Erie Extension Canal. It was constructed by some of the leading engineers of the state, Major David B. Douglass of the West Point engineers, W. Milnor Roberts, Charles T. Whippo, James Ferguson, and Sylvester Welch. The twenty-five-mile feeder ran from Meadville to Lake Conneaut and continued to Franklin on the Allegheny River.

The Beaver Division was divided into two parts, the Conneaut Division from Beaver to Conneaut Lake and the Shenango Division to Erie, which was completed in 1844 by the Erie Canal Company. Together they made a 136-mile route from the Ohio River to Lake Erie. This was the first part of the state canals to be disposed of, given to the Erie Canal Company in 1845. Although these western laterals were unprofitable, they contributed to the growth of the coal and iron industry in the northwestern part of the state. The first Lake Superior iron ore came down the Erie Extension Canal in 1853 and helped to bring the modern blast furnace to western Pennsylvania. Many western Pennsylvania towns and cities owe their origins to these canals. Pittsburgh grew from the canal trade and passenger traffic from northwestern Pennsylvania. Moreover, other lateral canals connected the Pennsylvania canals to the Ohio and Erie Canal across the state line, making a small network similar to that in eastern Pennsylvania.[45]

Taken as a whole, the Pennsylvania canal system was extensive and complex. It was almost more than the state could administer and operate. The canals were placed under the Board of Canal Commissioners, which was, as Louis Hartz has written, "from beginning to end, a strange and confusing thing."[46] The appointment of the board was "chaotic," there was fear that it would exercise excessive centralized control, the legislature retained a large sphere of power over the canals, and much local control was exercised by the engineers. By the time of their sale in 1857, the public works had cost $101,611,234 for construction expenditures and interest. Receipts from their operation and sale totaled $43,786,553, making a net loss of $57,824,681.[47] Yet the Pennsylvania canals helped to create towns and cities, carried salt and agricultural produce, and were almost indispensable for the establishment of the Pennsylvania coal and iron industry. They helped to make Pennsylvania an industrial state.[48]

Nowhere is this achievement more evident than in the anthracite canals of north-central and northeastern Pennsylvania. They did what canals did best: carried bulky, heavy cargoes to urban markets. A network of coal-carrying canals in Pennsylvania interlocked with each other and connected to canals in Delaware, New York, and New Jersey, over which coal could be transported to the New Jersey tidewater and New York City.

Three major anthracite coal fields stretched from Harrisburg to the northeastern corner of the state. The Schuylkill River reached up into the southern anthracite field. The Lehigh River passed close to the eastern end of the southern field and tapped the center of the middle anthracite field around Hazleton, and the most northeasterly field, the Wyoming, was traversed by the North Branch of the Susquehanna River and the Wyoming and Lackawanna valleys. It was these fields that were penetrated by the North Branch Canal and its Wyoming Division. An especially high-quality coal lay just below Wilkes-Barre and above Pittston.[49] Moreover, a field of bituminous coal lay along the West Branch of the Susquehanna River and extended to the Juniata, Allegheny, and Monongahela river valleys. In all, a third of the land of Pennsylvania overlay its coal fields.

The exploitation of the coal fields was a primary force in determining the course of the Pennsylvania canal system. Most immediately, the canals were designed to carry this coal to urban markets. But the canals were to be agents of economic development and instruments of political policy as well.[50] The coal fields were important in the logrolling politics that produced the Mainline Canal and help explain the decision to include the northern Susquehanna branch canals in the initial Mainline legislation. From the beginning, the development of the coal fields was a cardinal element in the movement for internal improvements in Pennsylvania. Great Britain was seen as a model for economic growth, and the analogy of British coal canals was readily apparent. Development of coal would also mean development of an iron industry, and the same canals could carry lumber, salt, gypsum, and agricultural products. This interest was evident when the Schuylkill Navigation Company was chartered in 1815 (a decade after the pioneer Schuylkill canal companies had ceased operations), and three years later an expanded program was begun to improve the Lehigh River navigation system.[51] Philadelphians' efforts to improve the Schuylkill and Lehigh valley routes to tap the coal fields rivaled, if they did not exceed, their interest in an east-west waterway.

Northern counties provided critical votes needed for the Mainline bill to become law, over the opposition of southern counties, whose interests were tied to Baltimore.[52] Appeals to the coal interests were evident in the arguments of the canal commissioners, in the Harrisburg convention of 1825, and in the legislature that enacted the Mainline system in 1826. Revenue from carrying coal was needed to pay for canal construction, which may explain why the canal act of 1827 included the West Branch Canal on the Susquehanna to Bald Eagle Creek and the North Branch Canal to the New York State line. The same act included the Delaware Division Canal from Bristol to Easton, important for delivering coal. The north-south direction of the upper Susquehanna coal trade was especially evident in the North Branch Canal.

The North Branch Canal was the first of three sections of canals that ran along the North Branch of the Susquehanna River.[53] It began at Shamokin Dam at Northumberland and extended fifty-five miles up the river to Nanticoke Dam, which supplied water for the canal. This section was completed in 1830, and its chief engineer was Charles T. Whippo, who had worked on the Erie Canal.

Above the Nanticoke Dam, the canal continued as the Wyoming Division for seventeen miles, passed through Wilkes-Barre to the mouth of the Lackawanna River at Pittston, and came to dominate the development of the Wyoming Valley. This was the heart of the upper anthracite coal field, but the rich agricultural valley of the North Branch produced grain, pork, and other products as well. On the Wyoming Division boats used a five-mile slackwater navigation behind the Nanticoke Dam and then continued through a separate canal to Pittston that was completed in 1834. This second section allowed navigation from the Susquehanna Division below Northumberland up the North Branch Canal and through the Wyoming Valley, following the river and penetrating the mountains of northern Pennsylvania. Coal or agricultural products could descend along the Susquehanna to Chesapeake Bay or be carried over the Union Canal to Philadelphia.

The third section on the North Branch of the Susquehanna, which was begun in 1836, was the North Branch Extension Canal. This ill-starred venture followed the river northwest from Pittston to the New York State line just above Athens, making another ninety-four miles of canal. Construction here was divided into the thirty-five-mile Tioga line along the upper river and the fifty-four-mile Tunkhannock line, which connected with the Wyoming Division at Pittston. Three river dams were required, one at Athens on the Chemung River, one at Towanda, and one at Horse Race rifts below Mehoopany, where the canal made a five-mile loop around a mountainous horseshoe curve. Sporadic construction on this almost impossible route approached futility and was stretched over two decades. A semblance of completion was not reached until 1856.

At the state line a private company added the eighteen-mile Junction Canal, built mostly in New York, to connect the North Branch route with the New York lateral canals reaching south from the Erie Canal. The Junction Canal reached the Chemung Extension Canal at Elmira and allowed passage through Seneca Lake to the Erie Canal. Or North Branch boats could use the Chenango Canal going eastward to Binghamton and north to the Erie Canal at Utica.

The effort to build a canal along the northernmost reaches of the Susquehanna Valley on the North Branch Extension had a troubled history. Expanding production from the anthracite mines, the almost incredible demand for coal in the 1840s and 1850s, and beckoning New York markets kept up a clamorous pressure for progress. But the Panic of 1837 dried up state funds, and work was suspended in 1841 after more than $2.4 million had been spent on the extension alone.[54] The North Branch Extension Canal was sold to a private company, which failed in spite of great expenditure, and construction was resumed by the state in 1849. Work was delayed by terrible terrain, porous soil, poor construction, changes in engineers, and frequent floods. The dams required building and rebuilding, twenty aqueducts were needed, and leaking banks lowered the water level in the canal. Still, by 1853 two boats from Pittston arrived on the Erie Canal loaded with anthracite coal, having traveled the North Branch Canal, the Junction Canal to Elmira, and Seneca Lake en route to Buffalo. By 1858 some thirty-eight thousand tons of Luzerne County coal had been sent to New York, and salt and gypsum moved south toward the other segments of the Susquehanna canals.[55]

Most of the coal in the upper anthracite field carried on the north branch route went south. In 1835 ten thousand tons of coal passed Berwick, and the 521 boats that cleared that port also carried agricultural products, merchandise, and lumber. Up-canal traffic brought dry goods, furniture, and other merchandise from Philadelphia and other eastern cities. In the 1840s the development of the anthracite iron industry led to increased coal shipments. The boom years came in the 1850s; in 1855 the North Branch Canal system carried more than five hundred thousand tons of coal. Tolls at Berwick rose from $33,000 in 1843 to $108,000 in 1853 and $165,000 in 1863.[56]

Packets on the North Branch Canal were built as early as 1832 and ran from Wilkes-Barre to Northumberland in a day at a fare of $3.50. They ran day and night, stopping at Berwick, Bloomsburg, Catawissa, and Danville on the way to Northumberland, offered food on board, and were typical of the packets of the Canal Era. Passengers continued on down the Susquehanna and Juniata divisions to Columbia, where they could take the railroad to Philadelphia.

A personal glimpse of the mixed experience on the North Branch Division, so important to the upper Susquehanna Valley and so frustratingly pursued, is found in the papers of the principal engineer, James D. Harris, appointed in 1836 at a salary of $2,500. Harris was assigned to the near impossible North Branch Extension Canal between Lackawanna and the state line, and he was in charge of maintenance on the West Branch and Susquehanna divisions as well. The latter divisions were no small task considering the constant problems of leakage, floods, changes in canal lines, and precarious state financing.

Harris's youthful contributions to the Western Division of the Mainline beginning in 1826 have already been mentioned. Hubertis M. Cummings has examined his papers and describes the pressures under which Harris worked:

> He was working in proximity to North Branch Susquehanna river folk who wanted a canal which should both connect them with waterways in New York State and with markets in Philadelphia and Baltimore. He had to weigh the commercial ambitions of Pennsylvanians from Athens to Towanda to Tunkhannock to Wilkes-Barre, observant of what editors of small-town newspapers from the New York State line to Northumberland thought, observant of miners' and manufacturers' importunities—as of legislators' wishes to respond to, or to create, those importunities. By the time he had made his examinations, determined his alternative locations, drawn his estimates of cost, summed up all in his official communications to the Canal Commissioners, and phrased his comments with due regard to his eventual reader, he had not only rounded his experiences as surveyor, engineer, man-of-affairs and diplomat, but achieved new masterdoms in mathematics.[57]

Party politics overrode whatever merit Harris displayed in meeting these challenges. He was identified with the anti-Jacksonians, now out of power, and he was dismissed as an engineer in 1839. Ironically, his position was taken by his close personal friend William B. Foster, Jr. (brother of the composer Stephen Foster). But Foster served from 1842 to 1849, when funds were cut off and the work was suspended. By the time work was resumed and the North Branch Extension Canal to the state line was finally finished in 1856, the canal found little use and this part of the North Branch route became one of the most expensive failures of the Pennsylvania system.

Before taking on the troubles of the North Branch Division, Harris had been appointed principal engineer of the West Branch Canal and put in charge of the Lycoming Line. The West Branch made a northern and western half-circle loop, separating from the Susquehanna Division and the North Branch at the major canal port of Northumberland. This canal followed the west fork of the Susquehanna River north to Williamsport and then turned west to Lock Haven. There, high in the mountains, it connected to the short Bald Eagle Cross Cut Canal. Twenty miles above Northumberland, the Muncy Dam furnished a water supply for the West Branch Canal, much as the Nanticoke Dam did on the North Branch Canal.

Harris faced all the problems he would meet again on the North Branch, locating the line, channeling the canal to match the nature of the river and its valley, struggling with the project of the Bald Eagle Dam, working with contractors, and even being sued by property holders for "forcible entry, trespass and damage" to their lands.[58] His personal papers show that his anti-Jacksonian politics led to his removal in 1835.

Boats on the West Branch Canal carried lumber and coal, iron ore and pig iron, merchandise, manufactures, and such agricultural products as wheat, hides, and whiskey. The Pennsylvania Canal Company dominated the trade, numbering its boats and maintaining stables at fifteen-mile intervals. A line of packet boats appeared in 1835, based in Harrisburg, and ran on both the Mainline and North Branch canals. A half-dozen more lines appeared, and daily packets ran from Northumberland to Williamsport and south to the Mainline at Duncan's Island. And at Muncy's Dam a canal hotel, accessible only by water, accommodated more than a hundred guests. These packets brought thousands of immigrants who setled in north-central Pennsylvania, and the West Branch Canal was traveled by political leaders who sought their votes.

James Buchanan came in 1844, Governor Ritner held a meeting of the canal commissioners at Lock Haven in 1838, Nicholas Biddle came in 1844, and later governor Andrew G. Curtin came by canal in 1852. For any political meeting of consequence, visitors came by canal and delegates departed by canal for meetings to the south.[59]

The stands of timber along the upper Susquehanna branches made this a major boat-building region of Pennsylvania. The Pennsylvania Canal Company built hundreds of boats at Espy on the North Branch, and on the West Branch the best-known yard was at Lewisburg just below Buffalo Creek. But dozens of builders, especially at Tioga, Wilkes-Barre, Northumberland, Sunbury, Selinsgrove, and Port Treverton, supplied boats for the Pennsylvania canals, ran dry docks, and attended the boats, which required constant repair.[60] Every spring, the lumber-laden rafts that went down the Susquehanna River to the Chesapeake were manned by crews that returned north by canal.

Many of these boats traveled far from the Pennsylvania coal fields, and some continued down the Susquehanna Valley toward Chesapeake Bay. Pennsylvania chartered the Susquehanna Canal Company in 1836 to build a canal from the Mainline Canal to Columbia, south along the west side of the Susquehanna, to join Maryland's canal built by the Tidewater Canal Company and terminating at Havre de Grace on the Chesapeake. The companies were informally amalgamated and operated together as the Susquehanna and Tidewater Canal. Pennsylvania's lower Susquehanna canal left the river at Wrightsville, where a dam was built to cross the river from Columbia. A double-deck towpath was added to the Columbia-Wrightsville bridge above the dam, and through canal passage on the lower Susquehanna was opened in 1840.[61] This canal, combined with the rebuilding of the North Branch Canal, created an increased outlet for Wyoming Valley coal.

But the completion of a canal that could carry Mainline boats south to the Chesapeake and nearby Baltimore allowed the diversion so feared by Philadelphians when they built the Union Canal and agitated for an east-west canal more than a decade earlier. And this new canal to the Chesapeake offered a larger capacity for traffic than was available on most other Pennsylvania canals. It was 50 feet wide and 6 feet deep, and its locks were 170 feet long and 17 feet wide, taking boats carrying 150 tons. Its operation and traffic will be described in Chapter 4.

The Delaware and Hudson Canal provided another diversion from the coal fields away from Pennsylvania to New York. Close by the upper branches of the Susquehanna River in the coal fields, the Delaware River offered an eastern outlet along the Pennsylvania–New York border. To exploit this water route, the Delaware and Hudson Canal was completed in 1828, reaching into the coal fields at Honesdale. The canal followed the Delaware River to Port Jervis in New York and then turned northeast to reach the Hudson River at Rondout, a distance of 108 miles. Beyond Honesdale a gravity railroad almost 17 miles long was built to the coal mines at Carbondale on the Lackawanna River, which flowed to the Susquehanna at Pittston and linked to the North Branch Canal in the Wyoming Valley. The path of the Delaware and Hudson was almost parallel to that of the Lehigh Canal, which would also reach the Delaware, and the Morris Canal, a short connection to New York.[62]

The Delaware and Hudson Canal Company was organized in 1823 by Maurice and William Wurtz of Philadelphia, who owned coal mines along the Lackawanna River and sought a market in New York City. Its president was Philip Hone, mayor of New York City, for whom the terminus at Honesdale was named. Once more New York engineers transferred their technological skills from the Erie Canal to a canal reaching into Pennsylvania. Benjamin Wright surveyed the line, and John B. Jervis soon became chief engineer.

Perhaps more than any other canal engineer, Jervis is an example of the transition from canal to railroad technology in the United States. His autobiography records his apprenticeship with Nathan S. Roberts and David S. Bates and, after eight years on the Erie Canal, his work as chief engineer on the Delaware and Hudson Canal.[63] Jervis developed inclined planes and put propellers on the cars to slow them on the grades. He called his system a "pneumatic convoy" based on "the resistance of air." Jervis brought the famous Stourbridge Lion steam engine from England when the canal was finished in 1829. It was the first locomotive to operate on a permanent track in America. The Lion, harbinger of a death knell for the canals of America, was brought through the canal to Honesdale; it was used briefly and then stored to rust because its eight tons were found to be too heavy for the tracks. Jervis would go on to work on other railroads and would design the Croton Aqueduct in New York City.

Maurice Wurtz worked with Jervis on the line. John Langdon Sullivan came from Massachusetts, where he had worked on the Middlesex Canal. Horatio Allen came from the Chesapeake and Delaware Canal. Among the other engineers, Porteous R. Root, James T. Clark, and James Archbald had served with Jervis on the Erie Canal. When Jervis left the Delaware and Hudson Canal, his successor was Russell F. Lord.

The Delaware and Hudson Canal is known as well for its two aqueducts over the Delaware River and the Lackawaxen Creek, for they were suspension bridge aqueducts designed by John A. Roebling. At Lackawaxen, where the canal crossed the Delaware River, Roebling's great suspension bridge aqueduct had four cable spans with cables eight and a half inches in diameter which carried a bridge 534 feet across the river, supporting coal boats weighing 130 tons.[64] His wire suspension bridge was accepted by Lord, who was then the chief engineer of the canal, partly because of the success of his suspension aqueduct over the Allegheny at Pittsburgh. Roebling built similar aqueducts on the Delaware and Hudson at Minisink Ford, at the Neversink River at Cuddebackville, and at High Falls near Ellenville in New York. All were finished in the spring of 1851, and they made possible dramatic increases in canal tonnage.

Initial coal shipments on the Delaware and Hudson Canal were modest, and it carried a general trade larger than that on any other coal canal. It did not show a profit for the company until after the Panic of 1837. But the development of the Pennsylvania mines and the growth of New York City led to an explosion of coal traffic, requiring successive enlargements from its original dimensions of twenty-eight feet by four to fifty by six feet in 1852. The first boats carried 20 tons of coal, then 30-ton boats were introduced, in 1845, 50-ton boats, and by 1852 130-ton boats could be used on the canal. In 1854 the Delaware and Hudson Canal carried more than a million tons of coal to market, paying tolls of more than $158,000, and coal tonnage peaked at almost 3 million tons carried to the Hudson in 1872.[65]

Still, the greater share of Pennsylvania anthracite coal found outlet to eastern markets through the great southeasterly river valleys, the Susquehanna below Northumberland and Columbia, the Schuylkill Navigation, and a combination of the Lehigh and the lower Delaware valleys.

The Schuylkill Valley was Philadelphia's first great link to the West and from it the Union Canal reached to the Susquehanna. But since the Schuylkill River rises in the southern anthracite field, it became preeminently a route for the export of coal.[66] The works of the Schuylkill Navigation Company, which was organized in 1815, touched the coal fields to the north at Pottsville and Port Carbon, met the Union Canal near Reading, and continued south past Norristown to the Delaware River below Philadelphia.

Along its 106-mile route, there was an almost equal proportion of canals and river improvements. From the time of the organization of the company in 1815, these were constantly being rebuilt, enlarged, or improved. A series of engineers, including Ariel Cooley, Thomas Oakes, Ephraim Beach, and George Duncan, supervised construction. There were 30 dams and 120 locks, and the 450-foot tunnel built in 1821 was the first transportation tunnel in the nation. Initial construction of the navigation cost $2.3 million. Navigation was opened to Pottsville in the coal fields in 1825, and although the company rented boats and charged tolls on its works, the navigation operated as a public highway. By 1845 the company had invested almost $4 million in the navigation, yet it was steadily profitable. Charles Ellet, Jr., was made president to supervise its enlargement between 1845 and 1847. In 1859 its gross annual income exceeded $1 million. Over time it became an increasingly complex system, changing the management of its boats, adjusting to the pressures of railroad competition in a freight-pooling agreement, and surviving a series of strikes. In 1834 some 570 coal boats made ten thousand passages through Reading. Most significantly, the tonnage on the Schuylkill Navigation grew from 32,000 tons in 1826 to 1,699,101 tons in 1859.[67]

Below the Schuylkill Navigation and the Delaware River, the turbulent Lehigh River also had its headwaters in the anthracite fields. In 1830, the year after the completion of the Delaware and Hudson Canal, the Lehigh Canal was built to carry anthracite coal down the Lehigh Valley to the Delaware at Easton. It was the work of the Lehigh Coal and Navigation Company, which sought a market in Philadelphia and New York for its coal from mines near Mauch Chunk.

The dominant figures in the company were Josiah White and Erskine Hazard. Together they improved on the earlier bear trap locks to create a slackwater navigation on the Lehigh River and built a gravity railroad from their mines to the river. They brought Canvass White from New York to survey the Lower Grand Section, and with him from New York came W. Milnor Roberts and Sylvester Welch. The latter soon called his eighteen-year-old brother, Ashbel Welch, to assist him. Yet White and Hazard moved radically away from the model of the Erie Canal. Their grand plan was closer to a ship canal using dams and slackwater rather than a series of lift locks. When the section from Mauch Chunk downstream to Easton was finished in 1829, its channel was sixty feet wide and six feet deep, and around the dams or other obstructions were forty-eight locks, one hundred feet by twenty-two feet, to carry boats of 150 tons.[68] The company then extended the canal twenty-six miles farther upstream from Mauch Chunk to White Haven, called the Grand Upper Division, and improvements with bear trap locks extended navigation still another twelve miles above White Haven to Stoddardsville.

After Canvass White died in 1834, Edwin A. Douglass worked on the upper section of the canal. He used some of Josiah White's innovations in dam construction and designed the highest lift lock constructed in America to that time, with a lift of thirty feet. Douglass had surveyed on the Western Division of the Mainline and resigned when his salary was reduced from $3,000 to $2,000 a year. And typical of the close personal relationship among the engineers of that time, he had been working with Solomon W. Roberts, who was the nephew of Josiah White.

Josiah White was one of the most remarkable figures of the Canal Era. Born in New Jersey and apprenticed at fifteen, he made a sizable fortune in the hardware business only to lose the business in a disastrous fire. Rescuing his finances by an agreement to pump water from his dam on the Schuylkill, he turned to the mines of the Lehigh Valley and moved to Mauch Chunk. He organized the Lehigh Coal and Navigation Company and devoted his energies to mining anthracite and improving the transportation of the valley. A devout Quaker, he had great integrity. Like others in the canal ventures of the era, he saw his company's interests as part of the economic development of a region.

Though White had no formal training, his inventive mind produced many innovations. Working with Hazard, he helped to discover how to burn anthracite coal and exploit its possibilities as a fuel. His ingenuity was evident in his designs for dams, iron boats, mechanical coal loaders, his Switchback Railroad for returning cars to his mines, and the nine-mile gravity railroad from the Summit Mines to the Lehigh River. Pennsylvania appointed him to its canal commission, and he personally supervised the rebuilding of the Delaware Division Canal. Until he died in 1850, the Lehigh Canal was in most respects his personal achievement.[69]

Like the coal tonnage on the Delaware and Hudson, that on the Lehigh Canal increased geometrically, from 225 tons in 1840 to 1.2 million tons in 1855.[70] Even when the Lehigh Valley Railroad was built parallel to the canal in 1856, canal tonnage continued to rise, and the company was profitable until after the Civil War. The company's coal and canal activities stimulated other developments, especially in the Lehigh Valley iron industry, which built on White's experiments in producing anthracite iron and his construction of America's first wire rope factory in 1848 at Mauch Chunk. There was also a tannery, and the use of Lehigh stone for cement laid the foundations for the Portland Cement Company. The company's activities contributed to the rise of cities such as Easton, Mauch Chunk, and White Haven. Alfred D. Chandler has written that the coal shipments arriving at the Delaware by the Lehigh navigation signaled the coming of industrialization in America.[71]

Coal traffic followed the full Lehigh navigation from White Haven through Mauch Chunk to Easton, there to take the Delaware Division Canal and the Delaware River to Philadelphia or to traverse the Morris Canal to New York. By 1840 more than a million tons of coal went down the Lehigh gorge by canal, a tonnage that was more than doubled by 1852. With the canal enlarged to allow boats of a hundred tons or more, traffic peaked in 1860. That year two thousand boats carried more than 2.3 million tons of coal and other products.[72] Floating amid this crowded traffic were the packet boats that carried passengers from Easton to Mauch Chunk, and when the Delaware Division Canal was opened in 1832 passenger service was extended to Philadelphia. Such tonnages on the Lehigh Canal reaching Easton give some sense of the vast coal traffic reaching Easton, which could be sent in two directions: south over the Delaware Division Canal to Trenton and Philadelphia, or northeast over the Morris Canal in New Jersey to New York City.

The Delaware Division Canal, a virtual extension of the Lehigh Canal, was built by the state and followed the Delaware River sixty miles from New Hope past Trenton to Bristol on the Delaware, where the river became navigable to Philadelphia. In contrast to the Lehigh, it was forty feet wide and five feet deep, but it was even more limited by its small locks, ninety-five by eleven feet, which reflected a political conflict with Senator Peter Ihrie of Easton, who opposed the Lehigh Coal and Navigation Company's plan for a larger canal.[73] Such small locks allowed boats only ten and a half feet wide, so that through boats could not take advantage of the twenty-two-foot-wide locks on the Lehigh Canal. Moreover, the locks and the nine aqueducts were so poorly finished that they all had to be rebuilt when Josiah White took charge of the work. Before White's improvements could be finished, the state sold the canal to the Sunbury and Erie Railroad in 1857, which immediately leased it to the Lehigh Coal and Navigation Company.[74]

In spite of devastating floods in 1841 and 1843, traffic grew steadily until some 8,700 boats came down the Delaware Division Canal in 1849, carrying not only coal but lumber, grain, flour, limestone, and iron. At New Hope an outlet lock into the Delaware River allowed boats to cross the river and enter the Delaware and Raritan Canal. A long navigable feeder of the Delaware and Raritan Canal on the north bank of the Delaware took boats into Trenton and north through New Jersey. Before it was sold, the Delaware Division Canal was the most profitable to the state of any canal in the system.[75]

Instead of following the route of the Delaware Division Canal from Easton southeast to the Delaware River at Bristol, canal shippers could use the Morris Canal to New Jersey. The Morris Canal left the Delaware and traversed New Jersey in a great arc to the north before it followed the Passaic River to tidewater at Newark Bay. Canal boats could cross the Delaware at Easton and enter the Morris Canal at Phillipsburg, go northeast as far as Paterson, and then turn south to Newark. The canal reached a final terminus at Jersey City on the Hudson River opposite New York. It was one of the two great canals of New Jersey.

The Morris Canal was the work of the Morris Canal and Banking Company, chartered under complex circumstances in 1824. Only 55 miles separated Easton, Pennsylvania, from Newark, but the Morris Canal almost doubled this distance in a route of 102 miles through rugged country. It climbed 760 feet for its water supply at the summit level at Lake Hopatcong, then fell 914 feet to Newark Bay. This 1,674-foot rise and fall in 90 miles was second only to the rise and fall on the Pennsylvania Mainline Canal. Twenty-three lift locks were used to surmount this remarkable elevation, aided by the twenty-three inclined planes designed initially by James Renwick. Without the inclined planes, some two hundred locks would have been required.[76] Renwick, who taught at Columbia University, had been trained in engineering in England and later wrote a classic biography of De Witt Clinton. Ephraim Beach, an Erie Canal engineer, became chief engineer on the Morris Canal in 1825 and remained until 1836. De Witt Clinton and Benjamin Wright came from New York to examine the route. They were followed by Joseph G. Swift, who had headed West Point, and the army engineers General Simon Bernard and Colonel Joseph G. Totten.

A succession of engineers invented new designs for the planes.

Ephraim Morris made the change from summit planes to lock planes, and David Bates Douglass made further refinements when he became chief engineer of the inclined planes. William H. Talcott introduced Roebling's twisted wire cables, which turned on a cable drum, to pull boats over the planes. But innovations were piecemeal and several different types of planes were in use at the same time.

Canal boats were floated onto cradles which were drawn on tracks over the inclined planes, pulled by cables powered by deep underground turbines, using water from the canal. Similar to the practice in Pennsylvania, rising boat cradles were balanced by the weight of descending boat cradles. Hinged section boats were needed after 1845 to get over the hump on the incline at the higher canal level. These complicated works cost $2.1 million, and the canal was completed to Paterson and Newark in 1831, with an extension to Jersey City in 1836.[77]

For all the ingenuity of its design, the Morris Canal was initially only thirty-one feet wide, and its seventy-five-foot-by-nine-foot locks would take boats of only twenty-five tons. Like the Delaware Division Canal, it was too small for the wider boats used on the Lehigh Canal. By 1845 its coal tonnage was only 58,259, but by 1850 that figure had risen to 238,682 and by 1860 to 707,631.[78] And this hundred-mile route usually consumed five days going from Phillipsburg to Newark.

In its earlier years, cargoes were partly agricultural, with potatoes, corn, sugar beets, and flour bound for Newark and New York, as well as sandstone blocks from quarries near Paterson. But it was primarily a coal-carrying canal, taking coal over its long route for about $2.25 a ton in the early 1830s.[79] The Morris Canal was an anomaly: a coal-carrying canal that could not accommodate boats over twenty-five tons, with narrow locks that could not take the larger boats from the Lehigh Canal, where its coal traffic originated. And as was common in the Canal Era, it operated in the midst of constant rebuilding. The locks and planes were enlarged in 1840-41 to take boats of fifty-four tons, and after remodeling begun in 1844, sixty-five- and seventy-ton section boats hinged in the middle came to the canal. The cost of the enlargement finished in 1860 was more than $6 million.[80]

But equally anomalous was the history of the origins of the Morris Canal, which were interlocked with the origins of New Jersey's other major canal, the Delaware and Raritan. No canal in New Jersey was built by the state, but as H. Jerome Cranmer has written, the New Jersey canals were built and operated in a climate of intense and prolonged

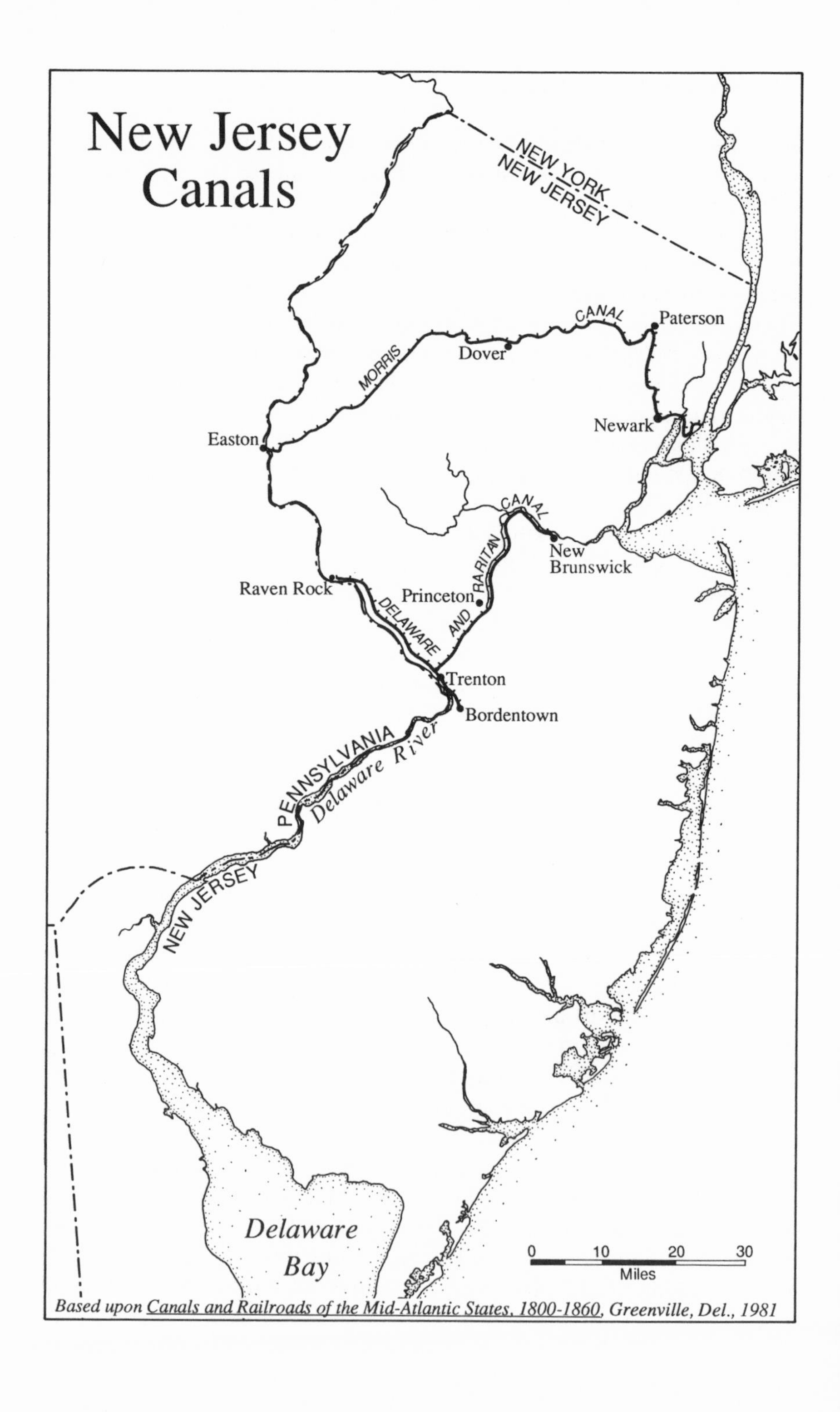
New Jersey Canals
NEW YORK
NEW JERSEY
Paterson
CANAL
MORRIS
Dover
Newark
Easton
CANAL
RARITAN
New Brunswick
Raven Rock
Princeton
DELAWARE
AND
Trenton
Bordentown
PENNSYLVANIA
Delaware River
NEW JERSEY
Delaware Bay
0
10
20
30
Miles
Based upon Canals and Railroads of the Mid-Atlantic States, 1800-1860, Greenville, Del., 1981

legislative debate.[81] In addition, New Jersey's canal history was more directly tied to both banking and railroad development than that of any other canal state.

Initially, the Morris Canal grew out of the public advocacy of George P. McCulloch of Morristown, who was interested in the agricultural development of northern New Jersey. But coal was also needed to revive the iron industry there, as many forges had been abandoned for lack of fuel. The New Jersey legislature appointed McCulloch to a commission to survey a canal route in 1822, its report was favorable, and in the legislative session of 1824 the canal dominated debate.

But northern New Jersey interests were soon challenged by John Rutherford, who represented interests backing the Delaware and Raritan Canal in the southern part of the state. Thus the state canal plan was defeated by sectional opposition, as was a "mammoth plan" by which a private company would construct both canals. Finally, the legislature chartered the Morris Canal and Banking Company, capitalized at $1 million with unusual banking privileges. These banking privileges were the price of creating a company that would perform what was deemed a public task, which could not be done by the state because of sectional conflict. William Bayard, the wealthy New Yorker, became the company's first president, and Cadwallader Colden, former mayor of New York, followed him in 1826. The company built the Morris Canal and in the bargain supported the charter of the Delaware and Raritan Canal Company in southern New Jersey.

Louis McLane, the former secretary of the treasury for Andrew Jackson, became president of the company in 1835 and the following year the directors leased the Morris Canal to the Little Schuylkill and Susquehanna Railroad company. Meanwhile, the Morris company expanded its banking and financial activities. Samuel Southard headed the company in 1837 and wrestled with its increasingly troubled finances. But when Edward Biddle, brother of Nicholas Biddle, became president of the company in 1838 the fortunes of the company took a downward spiral, burdened by embezzlement, fraud and mismanagement.

The speculative interests of New York dominated the Morris Canal and Banking Company and led it into wildcat banking, financial manipulation, and continual fluctuation in the company's stock. Such activities included a speculation in Indiana bonds in 1838 which had a disastrous impact on the Indiana canals. The company collapsed in 1841.

A reorganized Morris Canal and Banking Company took charge of the canal and began an ambitious enlargement program, which was under way from 1845 to 1861. The new company procured new boats, rebuilt the planes, and attempted to increase its share of the coal traffic coming east from the canals and railroads of Pennsylvania. In her detailed study of the Morris Canal, Barbara N. Kalata has traced the connections between the canal and a myriad of connecting railroads, which were part of the history of the Morris Canal beginning in the 1830s. Not only did the railroads enlarge the scope of the canal's activities, she found, but the canal was a "guiding force in the growth of those very railroads."[82]

In southern New Jersey, the Delaware and Raritan Canal demonstrated that canals and rails could be consolidated elements of an integrated transportation system. With its navigable feeder the Delaware and Raritan Canal had the shape of a giant "S." Its twenty-one-mile feeder ran from Raven Rock near New Hope beside the Delaware River south to Trenton. The canal itself began at Trenton and passed forty-four miles through southern New Jersey to meet the Raritan River, which it followed north to New Brunswick at tidewater. The canal turned northward from Trenton, and an extension was added from Trenton to return to the Delaware River downstream at Bordenton. This New Jersey canal, completed with its navigable feeder in 1838, became a reality in the trade-off between the northern and southern sections of the state, but it was a project that antedated the Morris Canal in public debate by many years. And its origins were as political as those of any canal in the Canal Era.[83]

William Penn advocated such a canal in 1676. John Pintard, who was closely associated with De Witt Clinton and the Erie Canal, wrote of it when he was a student at Princeton in 1774. The New Jersey Navigation Company made surveys in 1804, and Gallatin included the route in his report of 1808. In the early 1820s the Delaware and Raritan Canal Company was chartered, but its project was blocked by conflict with Pennsylvania over the diversion of water from the Delaware.

When the completion of the Erie Canal in New York heightened the prospects for state construction, it became part of the "mammoth plan" included with the Morris Canal until John Rutherford (who called this waterway the "Grand Trunk Canal of the United States") argued that it should not be combined with the Morris Canal project. Ironically, the separate company chartered in the compromise with northern canal interests could not proceed and was dissolved in 1828.[84] Again the state almost undertook the project, but this time the Camden and Amboy Railroad scheme stood in the way. When a new compromise between canal and railroad interests was reached, the Delaware and Raritan Canal Company and the Camden and Amboy Railroad Company were chartered together in 1830. Two years later, the combined companies were given a monopoly on the transport of passengers and cargo across New Jersey between Philadelphia and New York.[85]

The Joint Companies, as they have been called, fought off a group of New York, New Jersey, and Pennsylvania transportation interests in a bitter contest for public and legislative support. The Delaware and Raritan Canal Company spent large sums for legal fees, printing, and lobbying. When Chief Justice Roger B. Taney ruled against the monopoly that had been granted to the Joint Companies in 1832, the canal company argued that the combination was essential to its success. And when the canal company offered to sell the waterway to the state, the legislature declined the offer. Finally, in 1836, the Joint Companies agreed to a virtual consolidation with their transportation rivals, absorbing the railroad and turnpike companies involved and strengthening their combination against further attacks.[86]

The financing of the Delaware and Raritan Canal was primarily the work of Commodore Robert F. Stockton of Princeton. Stockton and his father-in-law, John Potter, bought half a million dollars of the company's securities. Nicholas Biddle and the Second Bank of the United States loaned more than a quarter million dollars to the company, and the company borrowed in Europe to meet the rest of the relatively low $1,175,000 cost of the waterway.[87]

This was the last canal constructed by Canvass White who died in 1834, the year the main canal was finished. He followed a route first surveyed by John Randel, Jr., as early as 1816, but White laid out a canal of unusual dimensions, 75 feet wide and seven feet deep. At the time, only the Chesapeake and Delaware Canal was larger. Its 44-mile route required only fifteen locks. Their size of 110 feet by 24 feet and its sixty pivot bridges allowed the canal to accommodate sloops and schooners. Most boats were pulled by mules and horses, but a steam packet operated in 1834, and steam-driven freighters regularly passed through the canal.

There was distinctive beauty in the aqueduct arch designed by the Princeton architect Charles Steadman. Steadman's Greek revival houses in Princeton were built of stone brought in by canal, and Princeton became the center of canal management. Yet the name of the engineer,

Ashbel Welch, is as closely associated with the Delaware and Raritan Canal as any in the history of this waterway. Welch was called by White from the Lehigh Canal, and he designed the twenty-three-mile navigable feeder along the northern bank of the Delaware, sixty feet wide and six feet deep and almost half as long as the Delaware and Raritan Canal itself. After White's death, Welch became chief engineer for the Joint Companies, a position he would hold for thirty-six years. In 1855 he doubled the length of the locks on the canal, and in a stroke of innovation he replaced the lock at Bordenton by building it away from the site, towing it on the canal, and sinking it into place.[88]

First called "Stockton's Folly," the Delaware and Raritan Canal was a conspicuous success. The Joint Companies had a virtual monopoly on the inland route between Philadelphia and New York, the two greatest mid-Atlantic ports, and some of their boats went on to New Haven and New London in Connecticut. Below Bordenton, where the canal connected to the Delaware, three canals entered that historic river. The Delaware Division Canal entered the Delaware at Bristol; the Schuylkill Navigation reached the Delaware at Philadelphia; and at the head of Delaware Bay, the Chesapeake and Delaware Canal cut across the peninsula to reach Chesapeake Bay and tapped the canal cargoes that had come down the Susquehanna and Tidewater Canal from Pennsylvania. The consolidated management of the Delaware and Raritan Canal and the Camden Railroad attracted investors from New York and New Jersey, and the canal earned its investors a steady return of 10 percent by 1860.[89]

Tonnage on the Delaware and Raritan was, of course, dominated by the coal carried on this web of canal connections, some down the Schuylkill Navigation into Philadelphia and the Delaware River, and some from the Lehigh Canal, down the Delaware Division Canal, crossing the river to enter the navigable feeder alongside the Delaware and finally to follow the Delaware and Raritan north from Trenton to New Brunswick and New York. The wider locks of the navigable feeder allowed the Delaware and Raritan to steal the coal trade from the Morris Canal. But the Delaware and Raritan also carried lumber, flour, grain, and merchandise. John A. Roebling moved his wire rope factory to Trenton close to the canal in 1850 and sent his famous cables out by water for installation on bridges, inclined planes, and aqueducts, many of which were on other canals. Coal tonnage totalled 1,283,265 in

1860.[90] The volume of this trade sometimes exceeded that of the Erie Canal.

H. Jerome Cranmer described the Delaware and Raritan as "exploitative" rather than developmental, a canal designed to exploit opportunities already existing. By contrast, in this interpretation, the Morris Canal was a developmental canal.[91]

The New Jersey canals and the canal between Chesapeake Bay and Delaware Bay created a network by which Pennsylvania trade could reach eastern markets. In northern Pennsylvania, branch canals connected to the lateral canals in New York, which reached the Erie Canal. The Delaware and Hudson took Pennsylvania coal across New York to the Hudson above New York City. To the south the Susquehanna and Tidewater reached tidewater on Chesapeake Bay at Havre de Grace, from which cargoes could be carried across the bay to the Chesapeake and Delaware Canal and then up the Delaware River to the New Jersey canals.

Nearer the center of the Pennsylvania coal fields, the Schuylkill Navigation reached from Port Carbon to Philadelphia, joined at Reading by the Union Canal. The Lehigh Canal reached the Delaware at Easton, from which coal could be shipped to Bristol on the Delaware River below Trenton by the Delaware Division Canal, or near New Hope could cross the river to the navigable feeder of the Delaware and Raritan Canal and then turn north to New Brunswick on the Raritan River. Coal arriving at Easton could also cross the Delaware River and enter the Morris Canal on a northern swing to Newark and Jersey City, a short distance from New York City.

Notwithstanding the losses on the Mainline or other segments of the Pennsylvania system, this was a notable canal and river network. Yet it was created by incremental steps, it was uneven in its design, and its operation was subject to constant competitive stresses. Each individual segment emerged out of unique origins, involving both state and private planning and often the work of powerful personalities. Only by frequent recourse to a map can one grasp the place of each segment in a comprehensive system.

The Pennsylvania canal system may well be judged a reckless, misguided failure, wasting colossal sums that might better have been spent on railroads. Yet the audacity of the total canal-building effort excites the imagination, and one can scarely comprehend the ways in which each canal segment contributed to the system as a whole.

4

Chesapeake and Southern Canals

While Pennsylvania and New Jersey sought to bring the Ohio Valley trade to Philadelphia and across New Jersey to New York City, rival efforts were under way to draw this trade to the Chesapeake and waterways in the South. They centered around the Chesapeake and Ohio Canal, the Baltimore and Ohio Railroad, and the James River and Kanawha Canal.

The Chesapeake and Ohio Canal project originated in plans by the Potomac Company to develop the Potomac River as the great gateway to the West. That company's tiny improvements only foreshadowed the design in the 1820s for a great canal that would follow the Potomac to Cumberland and then connect to the headwaters of the Monongahela River, opening a navigable route to the Ohio. Altogether such a route would extend northwestward 341 miles.

If any lesson was learned from the Potomac Company's experience it was that assistance from the national government was vital. Gallatin's report of 1808 had stated that the route merited national assistance, and the movement for the new Chesapeake and Ohio Canal was invigorated by the nationalism of the period after the War of 1812. The Chesapeake and Ohio Canal Convention in 1823 was followed by surveys by United States engineers. However much it was "palsied" (as John Quincy Adams put it) by a transition to "an Era of State Power," the nation that had rechartered a national bank and aided a National Road was becoming the chief promoter of a national canal. Walter Sanderlin in his history of this canal called it the "great national project."[1]

But the United States engineers estimated such a high cost for a canal to the Ohio, $22.3 million, that a new convention met to reconsider in 1826. President John Quincy Adams called on the New York engineers James Geddes and Nathan Roberts to make a report in 1827. Their estimates of only $4.5 million for a canal to the foothills of the mountains at Cumberland, Maryland, made construction of the great canal seem feasible, and the Chesapeake and Ohio Canal Company was organized in 1828. The United States Congress was the largest subscriber to the new company's stock, taking ten thousand shares and ultimately losing $1 million. The old Potomac Company deeded its rights to the new company. Maryland added half a million dollars and the cities of Washington, Georgetown, and Alexandria subscribed $1.5 million more. A final $600,000 came from private investors.[2]

On July 4, 1828, as a crowd of about two thousand looked on, John Quincy Adams broke ground for the canal at Little Falls, a few miles above Georgetown on the Maryland side opposite the old Potomac Company's locks, which still functioned. In an oft-repeated story, President Adams, who had made internal improvements one of the goals of his administration, warmed to his task. But the Maryland soil yielded as reluctantly to his spade as had Congress to most of his nationalistic economic programs. His spade struck a tree root, and three or four more thrusts made no better impression in the soil. For all his celebrated coldness of manner, the president took off his coat, drove his spade into the ground once more, and raised a shovelful of earth. "At which," he wrote in his diary, "a general shout burst forth from the surrounding multitude." The exertion, and "the anxiety," as he added, exhausted him so much that he "was disqualified for thought or action the remainder of the day." Still, he wrote that his casting off his coat "struck the fancy of the spectators more than all the flowers of rhetoric in my speech, and diverted their attention from the stammering and hesitation of a deficient memory."[3] Clearly, the popular outburst had pleased this scholarly president, and construction on this great national project began.

The president of the Chesapeake and Ohio Canal Company was Charles Fenton Mercer, congressman from northern Virginia. In Congress, Mercer was the chairman of the Committee on Roads and Canals. For the project he first called the "Union Canal," he secured the company's charter, won the congressional stock subscription, and guided the company's decisions. He made the canal his own, much as De Witt Clinton had done with the Erie Canal and as other forceful figures would do with other canals of the Canal Era. Mercer believed strongly in federal aid for canals, his personal friendship with President James Monroe had helped secure Monroe's support for the surveys by federal engineers in 1823, and he used all his political shrewdness to influence the crucial question of the location of the tidewater terminus of the Chesapeake and the Ohio Canal.[4]

Conflicts over the location of a terminus were common in canal

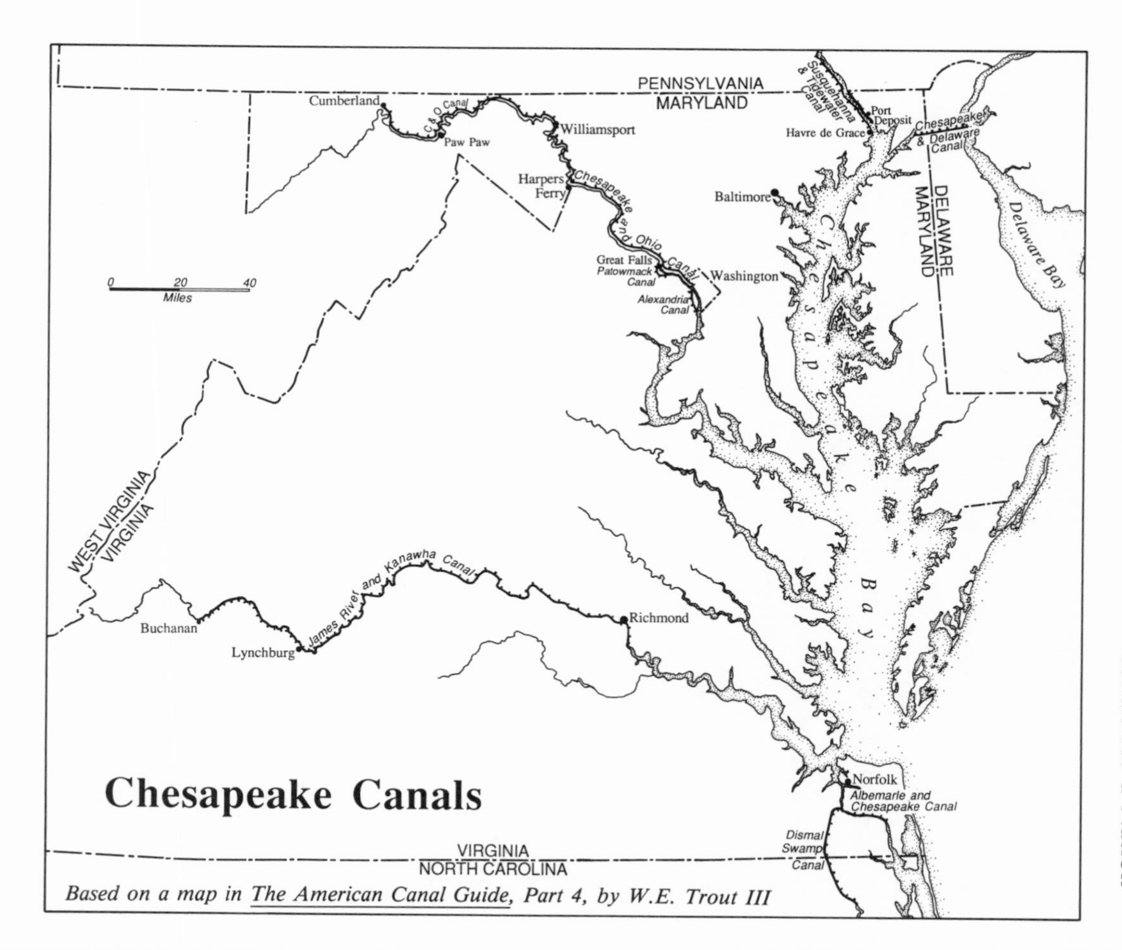

Chesapeake Canals

Based on a map in The American Canal Guide, Part 4, by W.E. Trout III

politics in the United States and occurred in New York, Pennsylvania, and New Jersey. When construction on the Chesapeake and Ohio began in 1828, work moved up the Potomac from Little Falls above Georgetown, and the question of the tidewater terminus was left unresolved for three years. Since the location of the terminus harbor was of importance not only for the technical viability of the canal but for obtaining political support for the entire canal project, the skillful hand of Congressman Mercer delayed final action until his own preferred termination in Washington could be agreed upon in a compromise decision.

In the engineering reports before 1828 the strongest recommendation had been for a terminus north of the Potomac at Baltimore on the Patapsco River, where there would be a deep-water harbor for oceangoing vessels. To reach the Patapsco the canal would leave the Potomac at Little Falls, pass through Georgetown and the District of Columbia, and proceed northeast toward Baltimore. But short of Baltimore, another deep-water terminus was possible by carrying the canal through Georgetown and the District of Columbia to the eastern branch of the Potomac at Bladensburg near the Navy Yard. And from there it would be only an additional twenty-five miles to the Patapsco at Elkridge, the southwestern entrance to Baltimore Harbor. Baltimore's connections to deep-water shipping and its trade and population gave it superior advantages over any terminus on the Potomac.

Tidewater on the Potomac meant three competing cities, whose rivalry entangled any decision about the terminus, while the manipulative hand of Congressman Mercer checked the competing interests until his own preference could be achieved. To placate the Marylanders, Mercer allowed the canal to be designated the Chesapeake and Ohio even while he worked for a terminus in the District of Columbia. Small boats coming down the locks around Little Falls entered the Potomac at Lock's Cove and were poled the two miles downriver to Georgetown, above which no ocean ships could go. This was some 108 miles from the mouth of the Potomac and seven or eight days of travel on the winding river. Georgetown foresaw a bright commercial future if the canal terminated there.[5]

But below Georgetown, Washington offered the prospect of a harbor in the mouth of Rock Creek on the boundary between Georgetown and Washington, and a terminus there was the ardent desire of Washington authorities. Nearby was the old city canal designed by Benjamin Latrobe and finished in 1815 from the Tiber Creek Estuary to the Navy Yard and the eastern branch of the Potomac. But another option would be to build an aqueduct across the river seven miles farther down the Potomac to carry the Chesapeake and Ohio Canal to Alexandria in Virginia, which then had the largest harbor on the Potomac.

After bitter and opportunistic rivalry, Mercer engineered his compromise: a thousand-foot promontory and dam would be built at the mouth of Rock Creek in Washington to make a harbor, which was finally done in 1831 when the first boat came down the canal, a packet, the *Charles Fenton Mercer.* Washington dredged out the old city canal to the deep water on the eastern branch and the Chesapeake and Ohio Company made a short canal to the Tiber estuary in 1833. But of far greater significance, an imposing aqueduct was built across the Potomac, and the canal continued from Georgetown parallel to the river and then went on to Alexandria. When this seven-mile Alexandria Canal was added in 1844, Alexandria became the major, if unofficial, terminus of the Chesapeake and Ohio Canal.

William M. Franklin had argued that the terminus question involved more than petty politics, personal ego, or even urban rivalry. It was a major flaw in the design of the Chesapeake and Ohio Canal, an ultimately fatal wound inflicted chiefly by a Virginia congressman who refused to follow the better wisdom of his engineers. In Franklin's view, the failure to develop an adequate Chesapeake outlet was also responsible for the failure of the company to complete the canal westward beyond Cumberland. When Rock Creek harbor silted up and the extension to the east branch proved unserviceable, there was no suitable harbor for the coal-carrying boats that finally came down the Chesapeake and Ohio Canal after it reached the District of Columbia in 1850. In the end, this lack killed the prospects for completing the canal westward to the Ohio River.[6]

If this was the consequence of Mercer's compromise, it is ironic because no one did more to bring the canal into existence than Mercer. He worked for the canal for more than a decade in Congress, in his search for financial contributions from the legislatures of Virginia, Maryland, and Pennsylvania and in his close supervision of his company's affairs. As a National Republican and a Whig he battled the Jackson administration to save his canal. As company president he fought the litigation of property owners who resisted the surrender of their lands for the canal, and he sought to prevent the encroachment of the Baltimore and Ohio Railroad.

This time the competing railroad did not lie unsuspected in the distant future; it was present from the beginning. The Baltimore and Ohio Railroad Company and the Chesapeake and Ohio Canal Company were chartered in Maryland on the same day in 1825. Moreover, Charles Carroll of Carollton, a signer of the Declaration of Independence, broke ground for the railroad on the same Fourth of July in 1828 when President Adams struck the celebrated root that made him shed his coat in his effort to turn the ground near Georgetown.

Baltimore merchants regarded the canal as a Virginia venture, and the city's commercial leaders decided to risk all on the newer, yet unproven, railroad. Their Maryland charter for a railroad to the Ohio River provided for a route to cross the Chesapeake and Ohio Canal at Point of Rocks in the Potomac Valley, where both canal and railroad would run along the river to Harpers Ferry. Thus the canal and the railroad were locked in controversy until 1832, fighting each other for the right-of-way on the north side of the Potomac. This conflict delayed canal construction above Point of Rocks, where the valley narrowed, for four years, when the courts decided for the prior rights of the canal.[7]

Pressure for money to construct the canal beyond Point of Rocks and to complete the eastern sections forced an appeal to both Virginia and Maryland, to which, ironically, only Maryland responded. While the Baltimore merchants backed their railroad, Maryland canal interests persuaded their state to take virtual control of the canal by a loan of $2 million in 1835, which was to be increased by $3 million more in 1836 and another $1 million in 1839. Thus in spite of the Baltimore-backed railroad, the Chesapeake and Ohio was virtually a Maryland canal.[8]

Construction began under a company-appointed Board of Engineers, which included Benjamin Wright, Nathan Roberts, and John Martineau, all of whom had worked on the Erie Canal. As Daniel Calhoun has pointed out, they were "organizational" engineers, more compliant to the demands of the company than the "independent" engineers from the army, who had made the earlier, higher estimates. They were joined by Charles Ellet, who worked on bridges and aqueducts before he left to study in France, after which he would return to work with Wright on the Chesapeake and Delaware Canal and also become involved in rivalry with John A. Roebling in their work on wire suspension bridges.[9]

The New York engineers faced challenges similar to those on the Erie Canal, as the Chesapeake and Ohio followed the north bank of the Potomac, much as the Erie Canal had followed the Mohawk Valley. As in New York, they supervised Irish labor, but in this case the company brought contract labor from Ireland and England, recruited by agents directed by the company president, Charles F. Mercer. They were paid $10 a month and promised meat three times a day and "a reasonable allowance of liquor." Although Mercer was an antislavery leader, his company used slaves to work on the canal. As on most canals, these laborers faced backbreaking work, and they endured the cholera epidemic of 1832. But their skills produced the special quality of stonework for which the Chesapeake and Ohio Canal is notable.[10]

The Chesapeake and Ohio Canal ran up the Potomac Valley to the coal mines at the foot of the Alleghenies at Cumberland but never reached its planned destination on the Ohio River. From tidewater to Georgetown, the canal passed the pioneer works of the Potomac Company at Little Falls and Great Falls, where the river flowed beneath high bluffs in a wild, rocky terrain. At Great Falls seven locks were clustered in a little more than a mile near the point where the impressive Cabin John Bridge would be built between 1859 and 1863. Just above the locks the first aqueduct was built at Seneca River, and in 1831 it became the first terminus of the canal. Two years later, the canal crossed the Monocacy River on the beautiful Monocacy Aqueduct and then was crowded by the high banks of the Potomac at Point of Rocks, forty-eight miles above Georgetown. At Harpers Ferry the canal passed beneath the high cliffs where the Potomac and the Shenandoah rivers break through the Blue Ridge. Above Harpers Ferry, the canal crossed Antietam Creek on an aqueduct and used a short slackwater navigation behind a river dam to reach Williamsport in 1834. About a hundred miles of canal were completed after work began in 1828, and eighty-five miles remained to reach Cumberland. "As far as it had been constructed," the leading historian of the canal has observed, "the Chesapeake and Ohio was without equal—the Erie Canal was likened to a mill race in comparison."[11]

But the company was plagued by insufficient funds, the enervating effects of the cholera epidemic, ethnic riots among the Irish, Germans, and other groups, and a general strike in 1838. The ethnic riots and the general strike were also marked by violence. Conflicts arose between Irish laborers from Cork, who had tried in a strike in 1834 to prevent Irish laborers from Longford from competing for canal jobs and now demanded a closed shop. In January 1838, three to four hundred Longfords met three hundred Corkonians near Williamsport; both sides were armed, and in a bloody conflict several men were killed and many were wounded. Work stopped on the entire canal line. Maryland sent in militia and appealed to President Andrew Jackson for federal troops. Jackson responded immediately. His secretary of war, John Eaton, had replaced Charles F. Mercer as president of the canal company, adding a personal relationship to Maryland's appeal for troops. Jackson's force kept peace throughout the winter and returned again when the laborers, numbering some three thousand on the whole line, struck for higher wages which the failing company could not pay. The exhausted company stopped work completely by 1842.[12] Coincidentally, 1842 was also the year of the stop and tax law that nearly halted New York State's enlargement of the Erie Canal.

The greatest engineering achievement on the canal was the Paw Paw tunnel, which was cut through the mountains above Williamsport. It was begun in 1836 to avoid five miles of loops in the river where close cliffs made canal construction almost impossible, and it was undertaken on the decision of the chief engineer, Charles B. Fisk. Ellwood Morris designed the tunnel, and the contract was awarded to Lee Montgomery, who had built the tunnel on the Union Canal. A channel 3,118 feet long was blasted through solid rock. It was 17 feet wide, with a 4-foot ledge for a towpath. The slate rock crumbled because of uneven overblasting, requiring a brick lining backed with fill many layers thick. Two sets of shafts were sunk to the tunnel for ventilation and to remove rocks. North from the tunnel a deep cut extended 890 feet farther to maintain the level. So great an undertaking brought about the financial failure of Montgomery as contractor, and several contractors later the tunnel, along with the rest of the canal, was completed to Cumberland by 1850.[13]

The little celebration ceremony in October 1850 at Cumberland was almost anticlimactic twenty-two years after the canal had begun with rosy expectations of reaching the Ohio and eight years since the Baltimore and Ohio Railroad had been finished to Cumberland. Boats carrying a hundred tons of coal each set out for tidewater. A Cumberland newspaper announced "the spring of enterprise at the entrepot of the Coal region." The editor anticipated "the arrival there of practical builders and boatmen, who have been taught their vocation on the New York Canals" and who would make "our Canal alive with boats" in a navigation season still open after winter had closed the Erie Canal.[14]

What the Cumberland editor celebrated was a Chesapeake and Ohio Canal now 184.5 miles long, carried to an elevation 609 feet above tidewater at Georgetown by seventy-four locks, each 90 to 100 feet long and 15 feet wide, crossing rivers and streams on eleven aqueducts. The longest aqueduct was the beautiful 438-foot structure spanning the Monocacy River, which rested on seven arches and was designed by Robert Leakie. Seven dams held back the river to supply a canal that has been described as one that "devoured water and was insatiable."[15] The Chesapeake and Ohio had a channel 60 feet wide and 6 feet deep. Above Harpers Ferry the canal was only 50 feet wide, and in the Paw Paw tunnel above Williamsport the channel narrowed to 17 feet for almost three-fourths of a mile. Only one boat at a time could go through the tunnel so there were often bottlenecks. A traveler passing through in 1859 wrote that the tunnel was "so narrow that the boat fitted it like a loose piston that crowded the water up into a wall in front of us, making hard pulling for the mules."[16] The boat moved so slowly that the short transit took nearly an hour.

Still, from the days of George Washington and the Potomac Company the project had been designed to tap the trans-Appalachian interior, and the company did not give up on its plans to build beyond Cumberland until well after the Civil War. The Chesapeake and Ohio Canal was largely a regional waterway, built at a cost of about $14 million, excluding interest and repairs, and it repaid less than 2 percent of its costs to the company that built it.[17] It was typical of the Canal Era, a canal that could not fulfill its original goals but nevertheless left an undeniable mark on the growth of towns and cities along its banks. Georgetown grew at the lower terminus, from Point of Rocks to Harpers Ferry the canal and railroad ran side by side, and above Harpers Ferry, Williamsport grew rapidly. Cumberland grew because of being a canal terminus.

Tonnage was always small: 60,000 in 1841, 100,000 in 1849. Traffic was dominated by flour and coal, and after 1850 coal was the chief cargo carried to tidewater, increasing the tonnage carried to 359,716 in 1859.[18] In 1851, 154 boats were in operation, and their number increased to 343 in 1856 in spite of disastrous floods in the 1850s. Yet the coal traffic was at the mercy of the Baltimore and Ohio

Railroad, which adjusted its rates to keep coal from being shipped on the canal.

Packet travel began when the first navigable stretch was completed and was recorded by Edward Coke on his journey on the first twenty-six miles in 1832. He noted the vivid contrast between the canal and the river that flowed beside it. "The canal flows smoothly and placidly along," he wrote, while the Potomac "is seen far beneath, rushing fiercely in a wild and tumultuous roar over a rough bed of rocks, and whirling along large trunks of trees with tremendous violence." Packets gave competition to the stages but were put out of service by the railroads. By 1859 another traveler riding the length of the canal on a line boat saw packets only for excursions.[19]

The Chesapeake and Ohio Canal bedeviled legislative politics in Maryland for decades, and its impact on Maryland's finances were nearly fatal. By 1841 the state was virtually insolvent. New money for the canal was raised by an act of 1844 that waived Maryland's liens on the canal and allowed the company to issue almost $2 million in new bonds, on which no interest would be paid after 1864. The canal's best years were in the 1870s. Five hundred boats were running in 1871 and carried almost a million tons of cargo. Disastrous floods struck again in 1878 and 1889, and revenues barely covered expenses.[20]

In 1899 most of the canal was sold. In that year, George Washington Ward, the first historian of the Chesapeake and Ohio Canal, surveyed its long history and concluded: "Such persistence deserved better results. Such heroic performance, even though attended almost uniformly with disaster, is unquestionably worthy of record upon the fair page of history."[21] But if the record on the page of history could not be fair, beauty has remained with the canal. Well-preserved masonry; broad, open towpaths; and the long sections of the Chesapeake and Ohio Canal still watered, like some of the canals of Pennsylvania and New Jersey, enable the present-day observer to recapture the flavor of the Canal Era.

The death grip on the fortunes of the Chesapeake and Ohio Canal held by the Baltimore and Ohio Railroad, as railroad competition controlled canal tonnage rates, amply rewarded the Baltimore interests who gambled on their unproven railroad in 1825. Yet there was great irony in Baltimore's transportation development. Baltimore was Maryland's primary seaport but did not benefit from the Chesapeake and Ohio Canal, of which Maryland was the major sponsor. Baltimore merchants still saw in the Susquehanna route an avenue to tap the Pennsylvania interior and southern New York, and they originated a project for a canal to parallel the Susquehanna River, which became a Pennsylvania venture in 1835. And when Pennsylvania pushed onward with the greatest canal on Chesapeake Bay, Baltimore opposed the project. This was the Chesapeake and Delaware Canal, which connected Chesapeake Bay above Baltimore with Delaware Bay and Philadelphia. It was unique in its tristate character, for it crossed Maryland and Delaware but its chief sponsors and beneficiaries were in Pennsylvania.[22]

The Chesapeake and Delaware Canal was only fourteen miles long, yet it joined the two greatest bays of the American Atlantic coast. Its origins went back to plans made in the seventeenth and eighteenth centuries, and the Chesapeake and Delaware Canal Company was incorporated in three states before 1802. It was first mentioned in the plans of Philadelphians who had tried unsuccessfully to bridge the Tulpehocken-Swatara route to the west and now sought to attract the trade of the Susquehanna away from Baltimore by a more southerly path and also gain a short water route to the Chesapeake. The Baltimore merchants were initially opposed, fully aware of the benefits certain to accrue to Pennsylvania.[23] The largest number of stockholders and the directors of the company were in Philadelphia; Baltimore merchants saw the Chesapeake and Delaware Canal as a rival to their control of the Susquehanna trade.

The first efforts of the company produced few results. Benjamin Latrobe was chief engineer and began construction, but investors flagged and funds ran out by 1805. Gallatin's report and public advocacy by the Philadelphians Mathew Carey and Joshua Gilpin brought new support so that construction could be resumed in 1824. Pennsylvania, Maryland, and Delaware together subscribed to $175,000 in stock.[24] But the crucial aid came from Congress with subscriptions of $300,000 in 1825 and an additional $150,000 in 1829. As its leading historian, Ralph D. Gray, has demonstrated, the Chesapeake and Delaware Canal was preeminently the "national waterway," built by a private company chartered in three states and aided by two congressional stock subscriptions.[25]

Ineffectual new surveys were made in 1822 by William Strickland, who had been trained by Latrobe. But with the renewal of the project and the completion of the Erie Canal in New York, the company promptly invited Benjamin Wright and John Randel, Jr., to lay out the route. Again organizational engineers, Wright and Randel, replaced the more individualistic Strickland.[26] Soon they were joined by Canvass White, who had left the Erie Canal to assist on the Pennsylvania canals. When the canal was finally built, it followed a more southerly route than had first been planned and was designed largely by Randel.[27]

The canal began at Chesapeake Bay, where Back Creek ran into the Elk River, which flowed into the Chesapeake. At the eastern terminus, the canal entered Delaware Bay at Newbold's Landing, forty-six miles south of Philadelphia. The canal was 66 feet wide and 10 feet deep, and thus almost a ship canal with only four locks. It was capped by a summit-level cut 280 feet wide at the top, 80 feet deep, and spanned by a bridge resting on a single arch. To the viewer of the time, the 240-foot bridge so high above the water was a spectacular sight.

But the New York engineers, Benjamin Wright and John Randel, disagreed about this construction, and the canal was completed amid bitter and expensive legal conflict. There were problems in digging through the marshes at the eastern end of the canal, where peat bogs and quicksand made it difficult to construct firm banks. The canal was finally finished in 1828, early in the Canal Era. Its cost was $2.25 million, but its large dimensions made the average cost per mile almost nine times that of the Erie Canal, which had been completed four years earlier.[28]

It was no longer necessary to take a water route of some six hundred miles from the Chesapeake Bay around the Delmarva Peninsula to the Delaware Bay, and Philadelphia had her long-sought easy access to the Susquehanna trade. Though Baltimore had more feared than desired the canal, the city now had a direct route to the Delaware. Freight boats on the canal paid the company from $4 to $12 a trip, and passengers made the canal passage in about two hours. Over the first three years, freight passages nearly tripled, from 2,379 in 1830 (carrying 61,500 tons and paying $24,658 in tolls) to 6,770 passages in 1833 (carrying 160,490 tons and paying $35,572 in tolls). Repairs and Randel's lawsuit against the company contributed to a temporary decline in traffic, but after 1837 the earlier rate of increase resumed until in 1850 there were 12,912 freight passages, constituting 361,640 tons and paying tolls of $198,364.[29]

Private companies operated both freight and passenger boats, often drawn by five or six horses hitched in tandem, and steamboats were seen briefly on the canal after 1835. After 1842 the Ericsson line of steamboats using the new Ericsson screw propeller dominated canal traffic.

Increasing lockages in the 1850s required more water, and in 1852 a giant steam-powered scoop wheel, 36 feet tall, lifted 130 tons of water a minute into the canal. Tonnage grew from 400,000 in 1852 to 600,000 in 1857; tolls averaged $220,000 a year, only one-fourth of which was required to meet expenses.[30]

The Chesapeake and Delaware Canal drew its trade from all over the Chesapeake region. Between one-fourth and one-half came from the Susquehanna Valley and central Pennsylvania as far north as Lycoming, Wilkes-Barre, and Mifflintown. Forest products, grain, and flour first dominated the eastbound trade, shipped from Havre de Grace at the mouth of the Susquehanna on the west side of the river or from Port Deposit a few miles upstream on the eastern bank of the Susquehanna. Coal became a major part of the southbound canal cargoes along the Susquehanna in the 1850s, moving toward Philadelphia. On the river itself arks and flatboats floated downstream in great numbers. And, of course, Baltimore sent its northbound trade through the Chesapeake and Delaware Canal.

The large dimensions of the Chesapeake and Delaware Canal allowed it to accommodate steam propeller boats, and there was little challenge from the railroads, which appeared only three years after the canal was completed. The New Castle and Frenchtown Railroad paralleled the canal in 1832, taking most of the passenger traffic, but canal freights continued to grow. In 1845, Delaware denied the canal company the right to charge tolls on passengers, and a decision by Chief Justice Roger B. Taney in the Supreme Court in 1847 upheld the right of passengers to travel toll-free on the canal as a public highway.[31]

The success of the Chesapeake and Delaware Canal in tapping the Susquehanna trade strengthened the movement in Baltimore for still another canal to the Chesapeake, this one along the Susquehanna itself, the Susquehanna and Tidewater Canal. It was chartered in both Pennsylvania and Maryland in 1835 and completed in 1840. In the beginning, it was a Baltimore project, and commissioners set out to survey a route from Conewago Falls above Columbia in Pennsylvania to Baltimore. They employed the peripatetic James Geddes as engineer and visited the Erie Canal themselves. Their return route from Cayuga Lake in New York along the Susquehanna in Pennsylvania to the Chesapeake impressed them with the vast region such a canal might draw upon to the advantage of Baltimore. But everything depended upon the willingness of Pennsylvania to join Maryland in permitting a canal to Columbia, where the Pennsylvania works ran along the Susquehanna, a project that was strongly opposed in Philadelphia. After years of delay, pressure from the Susquehanna counties induced the Pennsylvania legislature to relent. Over the opposition of many Philadelphians, in 1835 Pennsylvania chartered the company to build a canal over the twenty-six miles from Columbia to the Maryland line, and Maryland incorporated its Tidewater Canal Company for the remainder of the route to the Chesapeake. They were later combined as the Susquehanna and Tidewater Canal.[32]

The new canal was built on the opposite side of the Susquehanna from the older pioneer waterway designed by Latrobe to improve a short stretch of river navigation. It was only forty-five miles long, was 50 feet wide, and had locks 170 feet long, allowing boats carrying 150 tons. It terminated at Havre de Grace on Chesapeake Bay and cost $4.5 million.[33] Now Baltimore had her long-coveted canal route into Pennsylvania, and it appeared that one consequence would be to divert traffic from Pennsylvania's undersized Union Canal, which went from Middletown on the Susquehanna to Philadelphia. The Union Canal had already been bypassed by the Pennsylvania Mainline and the Columbia and Philadelphia Railroad, which was the eastern segment of the state works. Philadelphia might lose the small trade that still came over the Union Canal from the Susquehanna. In 1841, Pennsylvania authorized an enlargement of the Union Canal, which was begun in 1851. The bulk of the Susquehanna trade, it appeared, would go to Baltimore—but not yet. The Pennsylvanians ingeniously chartered a steam towboat company, towing boats from Havre de Grace at the mouth of the Susquehanna (and at Baltimore's doorstep) off to the northeast through the Chesapeake and Delaware Canal to Philadelphia. The company had secured special rates from the latter canal.[34]

The balance in the Chesapeake trade was delicate, much like that between the canals of Pennsylvania and New Jersey leading toward New York City. Although it was twenty miles farther from Havre de Grace to Philadelphia than to Baltimore, Philadelphia won the bulk of the Susquehanna trade. The number of boats towed to and from Havre de Grace through the Chesapeake and Delaware Canal to Philadelphia by the steam towboat service rose from 961 in 1841 to 4,806 in 1847. In 1850, Philadelphia attracted 2,576 boats from Havre de Grace, compared to 1,640 for Baltimore, and this increase was reflected in the fourfold rise in tolls on the Chesapeake and Delaware between 1840 and 1850. But in 1857 the balance shifted and Baltimore took 2,317 boats, outdrawing Philadelphia, which took 2,292.[35]

As was so often the case in the Canal Era, the Susquehanna and Tidewater Canal Company did not prosper from this growing traffic. Because of the high cost of construction, $80,000 per mile, and the depression of 1837 the original company was insolvent by 1842. Even the recovery of the early 1850s did not suffice, and the company suspended payment on its debts in 1859.[36]

While these first efforts to improve navigation from the Chesapeake north were under way, equally strenuous projects were afoot to link the Chesapeake by canals from tidewater Virginia through the mountains and to the West. The result was the James River and Kanawha Canal. At the urging of George Washington, the pioneering James River Company had been chartered in 1785, along with the Potomac Company, in a two-pronged effort to create a navigable waterway to the Ohio River. Only 33 miles separated the James River from the great Kawawha River and its tributaries, the New and the Greenbrier rivers, which flowed toward the Ohio Valley. The James River was navigable to Richmond and would be improved about 200 miles to the mountains. There the New and Greenbrier rivers might be improved to join the Kanawha, which was navigable almost a hundred miles more to the Ohio River at Point Pleasant, 485 miles west from Richmond. It is little wonder that Washington and others saw this as a great water route to the West.

In contrast to most other pioneer canals, the James River Company's first ventures were remarkably successful. The stock was oversubscribed in Richmond almost immediately. Under the company's first presidents, George Washington, Edmund Randolph, and Dr. William Foushee, short canals built for about seven miles around the falls above Richmond reached Westham in 1795. In Richmond, a three-mile canal terminated in the Great Basin, where thirteen wooden locks lowered boats to the James River. There were difficulties similar to those on other pioneer canals, but the company showed a profit by 1805, derived, it appeared, from high charges and inadequate maintenance. By 1805 the company had improved the James River navigation as far west as Crow's Ferry, 220 miles above Richmond, which led Albert Gallatin, in his report of 1808, to select this route as the easiest access to Appalachian coal. But when further improvements were desired which were beyond the company's ability to perform, the state of Virginia purchased the company's charter and made the canal a state enterprise.

The purchase was the result of the creation of commissions in 1811 and 1812 to reconsider a trans-Appalachian route through Virginia to the Ohio. The commission of 1812 was headed by John Marshall, who journeyed to the Kanawha and made a famous report advocating a through canal along the James, the New, and the Kanawha rivers, mountainous though the route might be. A Board of Public Works was created in 1816, along with a fund for internal improvements. The charter of the old James River Canal Company was purchased by the state, but the company itself was paid handsomely to perform the new work.[37]

The company placed its own interests ahead of those of the state; it did little and found itself unequal to the task. A new legislative act in 1823 removed the company from the work and made the project a state canal. A survey was done by Thomas Moore in 1819. He was assisted by a seventeen-year-old volunteer, Moncure Robinson, who was gaining his first experience in his canal- and railroad-building career. Another survey was made by Claudius Crozet, the state engineer who had also surveyed the Santee and Cooper Canal, and finally a definitive route was established by Benjamin Wright from New York. But Wright came into conflict with Crozet in 1824 when the work along the Kanawha was found to be far more expensive than had been estimated.[38] John Gamble, an engineer working on the waterway, wrote to Loammi Baldwin, Jr., that "sectional jealousies, want of public spirit and want of true political widsom" had almost "put a stop" to the project.[39] The company's canal was extended twenty-seven miles up the James River above Richmond to Maiden's Adventure Falls between 1823 and 1825 under the supervision of Moncure Robinson. Then in 1828 a seven-mile canal was added at the dangerous Balcony Falls, where the James River drops in a spectacular gorge through the Blue Ridge.

In 1828 Virginia held an internal improvements convention at Charlottesville, attended by men such as John Marshall, James Monroe, and Joseph C. Cabell. The participants were as distinguished as those who attended the Virginia constitutional convention that same year, and both conventions faced the same sectional conflicts between the tidewater and the West.

The 1828 internal improvements convention, meeting to find an answer to the challenge of a trans-Appalachian route to the Ohio, faced many of the same problems then being faced in Pennsylvania, Maryland, and Massachusetts. The convention resulted in a memorial to the legislature drafted by Monroe calling for canals along the entire James River–Kanawha River route, but when the memorial reached the legislature, few of its requests were approved. A turnpike across the mountains had already been constructed, and the canal won no new appropriations. As in Baltimore in 1828, the legislature was also considering a railroad proposal, which threatened any new canal proposal. In the surveys of 1828-29, Crozet favored a railroad over the mountains, but Benjamin Wright stayed with his proposed canal. When Wright was employed to survey the route again in 1831-32, he combined a canal to the mountains with a railroad over the rest of the route to the Ohio.

By this time the state had built works costing just over $1 million. Though this investment had been repaid, there was little prospect that the state would continue the work. Therefore, the backers of the canal project, especially John Marshall and Joseph C. Cabell, turned to private development and in 1832 won a charter for a new company, the James River and Kanawha Canal Company. This was the company that by 1840 would complete the canal 146 miles from Richmond to Lynchburg. It was called the First Grand Division. Work began on the more difficult Second Grand Division, and plans were to continue the canal to Pattonsburg and Covington on Jackson's River, where a railroad would run to the Kanawha. There a new canal would follow the river valley to the Ohio.

The new charter required capital of $5 million, which was finally raised over three years through the exertions of Marshall and Cabell. Shares were $100 each, and the state took thirty thousand, three-fifths of the stock. Richmond took almost six thousand shares, the Bank of Virginia took five thousand, Lynchburg one thousand, and private citizens made up the final five thousand.[40] If there was a guiding force behind the James River and Kanawha Canal, it was Joseph C. Cabell. That the funds were raised was a tribute to Cabell's persuasive belief that a trans-Appalachian route would do for Richmond what was being promised by similar projects for Baltimore, Philadelphia, New York, and Boston.

Construction was pushed vigorously in 1836 with fourteen hundred men at work, and the following year there were thirty-five hundred men on the line. In 1838 slaves were added to the work force until they made up two-thirds of the laborers on the canal.[41] Virginia now resembled northern states that were experiencing a similar burst of canal building. The canal was completed to Lynchburg, 146 miles above Richmond, and construction was under way on another 50 miles west of Buchanan. The old canal from Richmond to Maiden's Adventure was rebuilt to match the extension to Lynchburg, as was the 7-mile Blue Ridge Canal. But in 1842, the year when work stopped on most American canals, the finances of the company were so strained that the legislature prohibited further work until the company's obligations could be paid. To make matters worse, the most disastrous flood on the James River in fifty years caused more than a hundred breaks in the canal banks.

The canal's promoters were undaunted, and a convention at Lewisburg in 1844 sought state assistance. Cabell still worked for a canal over the entire route to the Ohio. The most that could be won in 1847 was a legislative appropriation that resulted in finishing the thirty-eight locks and 50 miles of canal to Buchanan in 1856, 196 miles west of Richmond. This would be the final terminus of the James River and Kanawha Canal. It was less than half the route originally projected, and a 208-mile road was built from Covington, Virginia, over the mountains to Point Pleasant on the Ohio River. The canal from Richmond to Buchanan had cost $8,259,184, about the same as the original cost of the Erie Canal in New York. It was defeated by the same mountainous barrier that had stopped the Chesapeake and Ohio Canal at Cumberland. Even so, in the 1850s another two-thirds of $1 million would be spent on the Third Grand Division to carry the canal 47 miles on to Covington. Fifteen miles of canal from Buchanan to Eagle Rock, through spectacular country, were put under contract, and a tunnel was begun before the project was abandoned.[42]

South of Richmond, at the tidewater end of the James River and Kanawha Canal, prospects for the canal were as bright as those on the rest of the canal were blighted. In the late 1840s a new tidewater connection was begun to replace the eastern terminus at the Great Basin in Richmond. Two turning basins and five stone locks replaced the original thirteen wooden locks to the river. The "great ship lock" at the river in 1854 opened navigation from the Atlantic and the James River into Richmond and the canal. Between 1855 and 1860 the number of vessels arriving and departing Richmond rose from 1,594 to 4,460, earning handsome dockage receipts for the company.[43]

Emphasis on the more numerous canals of the North and the Midwest may underrate the importance of the James River and Kanawha Canal. It was clearly a major canal of the nation. Its dimensions were larger than those of most northern canals, fifty feet wide and five feet deep; its ninety locks were one hundred feet by fifteen feet; and it included a dozen aqueducts and hundreds of other structures.[44]

Traffic reached impressive levels, with an annual tonnage of 231,032 in 1853 and 244,273 in 1860, when tolls reached $231,000. In 1860 down-canal traffic included nearly 18,000 hogsheads of tobacco, 53,046 boxes of manufactured tobacco, 695,388 bushels of wheat, 10,933 bushels of corn, 4,177 tons of pig iron, 21,305 tons of coal, 20,898 tons of stone, and almost 10,000 cords of wood. Up the canal went salt, plaster, castings, fish, nails, and guano. For 1859 the estimated value of goods on the canal was $21,658,000.[45] Close to four hundred boats used the canal regularly in 1854, including six regular packets. Unlike most other canals in the nation, the James River and Kanawha operated year-round. For the thirty-three-hour trip from Richmond to Lynchburg, passengers paid $7.50 before 1848, a fare that was lowered to $5.35 in the 1850s, resulting in increased passenger travel.[46] In 1863, General Stonewall Jackson's body was brought back from the battle of Chancellorsville on board a packet named the *John Marshall*.

Although the canal company was the largest corporation in the state and before 1860 carried more tonnage than did all the railroads in the state combined, the company was not profitable to its shareholders. High maintenance costs, especially on the Kanawha River improvements, were matched by high interest costs on the company's debt. The company owed $7 million by 1860 but received a little more than $5 million in tolls and water rents between 1835 and 1860.[47]

As was true elsewhere, the year 1854 was the turning point in the battle against competition from the railroads. Canal revenues grew steadily in the early 1850s but began to decline as railroad mileage increased. After 1837, as in New York, Pennsylvania, and Indiana, Virginia shifted its assistance from the canal to the railroads, although canal construction proceeded in the midst of debate about assisting railroads. In this controversy, many Virginians were alarmed at the Pennsylvania experience, which used both a canal and railroad route over the mountains. Yet an appeal published in late 1847 continued to nourish the futile hope for a through water route to the Ohio in Virginia.[48]

In Virginia politics, the Whigs more frequently supported the canal, but the Democrats were more frequently in power. Governor Henry A.

Wise was the last champion of the completion of the entire water route as he sought to bind western Virginia to the east amid rising tensions before the Civil War. An agreement with a French company to purchase the canal and complete it failed, and in 1861 the Virginia Canal Company was chartered to take over the James River and Kanawha Canal Company. With the outbreak of war, the venture collapsed.[49]

In 1880 the canal works were sold to the Richmond and Allegheny Railroad Company, which valued the easy grade along the James River to the Appalachians. Ultimately the James River branch of the Chesapeake and Ohio Railroad ran along the James, the Greenbrier, the New, and the Great Kanawaha rivers to the Ohio, often using the canal towpath for a roadbed. But for half a century and more, it was a route marked out for a canal.

These persistent efforts to draw western commerce to Richmond or Baltimore understandably obscure the simultaneous efforts in Norfolk and southeastern Virginia to improve an inland water route to the south, through the Great Dismal Swamp, to connect Chesapeake Bay with Albemarle Sound and below that to Pamlico Sound. At the center of these efforts was the Dismal Swamp Canal, which was among the pioneer canals of the 1790s, sponsored by George Washington and others, and which had become a ship canal by 1814. To draw the trade of Albemarle Sound toward Norfolk and Chesapeake Bay, the Dismal Swamp Canal began in North Carolina at Elizabeth City on the Pasquotank River, which flowed into the sound. The canal followed a northwesterly arc until it passed Lake Drummond, from which it took its water for lockage by a feeder canal. The arc of the Dismal Swamp Canal then turned northeasterly to Norfolk on the southern branch of the Elizabeth River, which connected to Chesapeake Bay, making a twenty-two-mile route through the desolate Dismal Swamp. From the beginning its progress was slow and spasmodic; but after the War of 1812 it entered an era of improvement and prosperity. By 1815, more than a million barrel staves and half a million shingles had been carried in flatboats through the waterway.[50]

On the long-standing precedent of Gallatin's endorsement in his report of 1808, Congress took six hundred shares in the Dismal Swamp Canal Company in 1826 and two hundred more in 1829, totaling $200,000. Virginia added fresh loans to its previous investment, and the new funds allowed extensive improvements. By 1828, $800,000 had been spent on the canal; the channel averaged forty feet in width and would take vessels with a draft of five and a half feet. Five new stone locks employing a cement that could be used underwater, made from Canvass White's "hydraulic lime" in New York, replaced the older wooden locks. Moreover, North Carolina added a six-mile branch canal to the Northwest River, which flowed into Currituck Sound, and the Roanoke Canal Company made a short canal, which increased the Roanoke River trade reaching Albemarle Sound. Fifty sloops, schooners, rafts, and lighters traversed the Dismal Swamp Canal in two weeks of June 1829. In 1833, northbound goods, led by lumber products, cotton, tobacco, and corn, were valued at $1,713,796; southbound goods, which included merchandise, salt, coffee, molasses, flour, and other products, were valued at $780,088. With such traffic the Dismal Swamp Canal Company prospered; tolls rose from $11,658 in 1829 to $34,059 in 1833. In the late 1830s some 350 vessels a year used the canal. Trade on the canal increased steadily, especially lumber products, but including 1.25 million bushels of corn in 1847. By 1854 the value of goods carried was more than $3.5 million. The stock purchased by the national government at $200,000 in 1828 was valued at $600,000 by 1861.[51] Clearly, the Dismal Swamp Canal was a profitable venture, even though its route passed through some of the most desolate and uninhabited swamplands on the Atlantic coast.

Why, then, was there a decades-long movement for another canal through the Dismal Swamp? What led to the building of the Albemarle and Chesapeake Canal, begun in 1855 and finished in 1859 near the end of the Canal Era? The route it took through the Dismal Swamp was an alternate choice in the 1730s and 1790s and a constant rival in the original plans for a coastal canal between Albemarle Sound and Chesapeake Bay. This route also enjoyed the hallowed status of endorsement in Gallatin's report of 1808. Even though it was not chosen, its prospects were renewed by every difficulty encountered on the Dismal Swamp Canal. Shoals and logs impeded traffic; the canal was too narrow for the sloops that could pass only at recesses built periodically into the sides of the channel; and there were problems in the twisting course of the canal where it joined the Pasquotank in the south and the Elizabeth River in the north. Vessels waited for days in the upper Pasquotank River, awaiting towage or detained by low water in the canal.

The Albemarle and Chesapeake Canal took a more easterly path through the Dismal Swamp, closer to the Atlantic shore. At its southern end it began at the North River, which like the Pasquotank flowed south

Moonlight on the Erie Canal, by J.W. Hill, 1832. Phelps Stokes Collection of the New York Public Library

Watercolor of boats on the Erie Canal, by J.W. Hill, 1830. Phelps Stokes Collection of the New York Public Library

View of Rochester. Reproduced from a large lithograph entitled *Rochester from the West, 1853*. University of Rochester Library.

Below, the aqueduct bridge at Rochester. From Cadwallader Colden's *Memoir . . . at the Celebration of the Completion of the New York Canals*

Stock Exchange Corner, Philadelphia, about 1845, where sectional canal packet boats loaded for Pittsburgh. Courtesy Hugh Moore Historical Park and Museums

Below, inclined plane number 6 on the Allegheny Portage Railroad. The cars and sectional boats were drawn up or down these planes by cable, powered by a stationary steam engine at the top. Courtesy Hugh Moore Historical Park and Museums

Weighing boats in the weigh lock on the Lehigh Canal. Courtesy Hugh Moore Historical Park and Museums

Chesapeake and Ohio Canal and Baltimore and Ohio Railroad, near Sandy Hook, Maryland. T.F. Hahn Collection

Aqueduct on the Chesapeake and Ohio Canal at the mouth of the Monocacy. *Harpers Weekly,* 1861. W.E. Trout III Collection

Below, James River and Kanawha Canal at Richmond. Courtesy Virginia State Library

Thomas Kelah Wharton's view of the Miami Canal, Dayton, Ohio, 1831-1832. Wallach Division of Art, Prints and Photographs, the New York Public Library, Astor, Lenox and Tilden Foundations

Miami Canal. *Ladies' Repository,* 1842. Courtesy Ohio Historical Society.

into Albemarle Sound, and ran almost straight north through the "North Carolina Cut." It crossed Currituck Sound and used the North Landing River in Virginia until the "Virginia Cut" turned straight west to reach the Elizabeth River, connecting to Norfolk and Chesapeake Bay. The Albemarle and Chesapeake Canal and the Dismal Swamp Canal thus entered the south branch of the Elizabeth River only a few miles from each other; southbound traffic on the river could take either canal to reach Albemarle Sound. The route of the Albemarle and Chesapeake Canal, sometimes called the Juniper Canal from the petrified juniper roots that had to be cut in its excavations, ran some seventy-five miles and is now part of the Intercoastal Waterway.

The history of this canal was as tangled as the juniper roots over which it passed. Like the Dismal Swamp Canal it was proposed as part of a great inland chain of waterways from New York to North Carolina. It was surveyed by Robert Fulton of New York, who reported to Albert Gallatin, by Hamilton Fulton for the North Carolina Board of Internal Improvements, and by T. L. Patterson for the city of Norfolk. The North Carolina Board of Internal Improvements had been created in 1819 by Archibald D. Murphey, a leading canal advocate, and the board hired the English engineer Hamilton Fulton to do its surveys. Fulton's reputation was so distinguished that when Georgia created a Board of Public Works in 1825 Governor Robert Troup selected him for chief engineer as the best in the South.

But the real force in the Albemarle and Chesapeake Canal Company was Marshall Parks of Norfolk, Virginia. His father had been manager of the Dismal Swamp Canal and knew its problems, and the younger Parks himself had been employed on the canal. Marshall Parks secured an amended Virginia charter for the Great Bridge Lumber and Canal Company in 1854 to cut a canal from the Elizabeth River to the North Landing River. He then proceeded to win a new charter in North Carolina "To Incorporate a Company to Construct a Ship Canal to Unite the Waters of Albemarle, Currituck, and Pamlico Sounds with Chesapeake Bay." Effectively this meant that the company in North Carolina could cut a canal from Currituck Sound to the North River, which emptied into Albemarle Sound. The Virginia and North Carolina companies were united in 1856 as the Albemarle and Chesapeake Canal Company.[52]

Only about nine miles of excavation were required on the Virginia Cut and about five more on the North Carolina Cut, and the newly available steam dredge technology made the canal possible. As Edmund

Ruffin wrote after visiting the project, "The surface of the swampy ground is, in many places, so nearly level with the water, and the earth so generally a quagmire of peat, and so full of dead roots and buried logs, under the water, and of living trees and roots over and at the surface, when but very little above water, that the difficulties of removing such obstructions are very great, and would be insuperable if by the use of ordinary utensils, with hand-labor."[53] The canal was sixty-one feet wide in both cuts, and an effort was made to maintain an eight-foot depth.

Seven of the new steam dredge machines were put into use in 1857. The invention of a "reversible-head" guard lock to counter the tidal effects in the Elizabeth River at Great Bridge allowed the canal to function. The Great Bridge guard lock was notable for its size, 220 feet by 40 feet (only the locks at the Sault in Michigan were larger), but this lock had to control surging currents caused either by the tidal rise of the Elizabeth River or the opposite effects of southerly winds on the sounds, which pushed up the level of water in the Virginia Cut. At either end of the guard lock, the water level might change so that vessels might have to be raised or lowered to enter or leave the cut instead of following established levels served by locks as on every other canal. Two alternative sets of miter gates were built at each end of the lock, with eight gates, 8 feet high and 25 feet wide. The lock tender selected the proper pairs so that their mitered points headed toward the higher level. The company described the lock as "probably the first lock ever constructed which would allow vessels to lock up or down either way."[54] One of the first vessels to use the canal was the *Enterprise* from North Carolina, which could carry ten thousand bushels of grain or six hundred bales of cotton.

The canal cost the company a little more then $1 million, which was raised by stock subscriptions, and it was almost immediately a financial success. It siphoned trade from the Dismal Swamp Canal because of its shorter route, higher water level, and accommodation for larger vessels. Traffic moved on it in 1859 even before it was finished. A thousand vessels used the canal in 1860 and more than twenty-five hundred in 1861.[55] Its cargoes were the same as those on the Dismal Swamp Canal, lumber products moving north in greatest volume; merchandise, flour, beef, pork, and guano made up most of the traffic going south. By 1859, then, there were two great canals linking the North Carolina waterways with the Chesapeake Bay in Virginia. They offered a protected, wide-water, sloop navigation alternative to the dangers of Cape Hatteras. And their inland route could be continued through the Chesapeake and Delaware Canal and the Delaware and Raritan Canal through New Jersey to New York.

As the Virginian Edmund Ruffin observed the great sixty-foot-long steam dredges at work on the Virginia Cut, he wrote that "the speedy and perfect completion of this great work . . . will give to North Carolina, for the first time a proper outlet for, and the proper use of her noble interior navigable waters."[56] This was also the goal of the North Carolina Board of Internal Improvements, which encouraged the chartering of companies to improve the Roanoke, the Tar, the Yadkin, the Neuse, and the New rivers. The board not only engaged Hamilton Fulton in England, it also brought Benjamin F. Baldwin, the son of Loammi Baldwin, south from Massachusetts to make canal surveys. South of Pamlico Sound to Beaufort Inlet, the three-mile Clubfoot and Harlow Creek Canal was cut in 1829 across the Carteret County Peninsula in hopes of making Beaufort a coastal metropolis.

But more significant was the Roanoke Canal, a nine-mile waterway completed in 1831 to bypass the falls of the Roanoke River at Weldon, which impeded that river as it ran from northeastern North Carolina into Albemarle Sound. Hamilton Fulton made the surveys in the 1820s, though his service in North Carolina became a source of controversy, and by 1834 the canal was completed. It boasted the Weldon Aqueduct, 110 feet long resting on a single arch, and it had two sets of two-lock combines. In 1835 it carried 10,646 barrels of flour and 6,877 hogsheads of tobacco, as well as large quantities of merchandise and other products. Tolls earned the company almost $9,000 in 1838, but competition from the railroad compelled the closing of the canal in 1859.[57] Yet its considerable traffic swelled the Albemarle trade going north over the Dismal Swamp Canal or the Albemarle and Chesapeake Canal.

Other North Carolinians sought to combine their canal plans with the emergent railroad. Joseph Caldwell, president of the state university at Chapel Hill, devised a canal-railroad plan in 1828 to develop Beaufort Harbor and link it with the Ohio and Mississippi rivers. Like Strickland in Pennsylvania, he traveled to Europe and returned convinced that railroads could help his state more than canals, and he tried to combine the two. A canal would reach from Beaufort on the coast to New Bern, and then a railroad might reach to Fayetteville, Rockingham, Asheville, Knoxville, and Nashville, where the Cumberland and Tennessee rivers would connect to the Ohio and the Mississippi. Caldwell calculated that the route to the Mississippi would be shorter than that of the Erie Canal, and it could be open year-round. His project never went beyond the planning stage, but it paralleled those devised in Richmond, Baltimore, Philadelphia, New York, and Boston. Moreover, it prepared the way for routes later selected for railroad construction.[58]

Meanwhile, the pioneer Santee Canal in South Carolina continued to operate and served as an example of the extent to which the deep South participated in the Canal Era. The Santee was not challenged by railroad competition until the 1840s, and it was not abandoned until 1858. Some $2 million were spent on the ten short canals begun in South Carolina in the 1820s. Some were abandoned after a decade, all by 1850. Canal technology moved from south to north as the contractor Robert Leckle went from South Carolina to become superintendent of masonry on the Chesapeake and Ohio Canal, accomplishing much of the masonry work that distinguished the latter waterway. John Wilson went from South Carolina to build the Columbia and Philadelphia Railroad and ran surveys for several Pennsylvania canals. The English engineer Charles B. Vignolls worked on canals in South Carolina before returning to Europe, where he wrote books on engineering.[59]

Canal-building activity in Georgia dispells any idea that the Canal Era was exclusive to the Northeast or Midwest. Gallatin proposed canals from Tennessee through northwest Georgia to the Atlantic, as well as south to the Gulf of Mexio. Governor David Mitchell worked for internal improvements in Georgia in 1815, sharing the nationalistic fervor at the end of the War of 1812. The completion of the Erie Canal in 1825 led Governor George M. Troup to propose canals from the Tennessee River to central Georgia, with links to the Savannah, Oconee, Ocmulgee, and Chatahoochee rivers. Milton Heath, the leading historian of these canals, noted that canal enthusiasm was at "fever pitch" in Georgia, where the same hope for western development was expressed that was evident in almost every Atlantic state: "Here seemed the possibility of creating the most likely parallel and rival of the Erie Canal anywhere in the United States."[60] In 1825 a Board of Public Works was created, and the next year, Georgia brought Hamilton Fulton from North Carolina to supervise its canals.

The board, with later governor Wilson Lumpkin as one of its members, conducted surveys of canal routes from the Tennessee River through the valley of Chickamauga Creek between Lookout Mountain and range of the Blue Ridge to central Georgia and to the Chatahoochee in the southwest. A controversy between Fulton and an assistant engineer, E.H. Burritt, led to Fulton's work in North Carolina being called into question. He was thought to be arrogant and pretentious, and he was accused of plagiarizing from the *Cyclopaedia* by Abraham Rees in an earlier report to the legislature. As a result, he was dismissed as chief engineer at the end of 1826.[61]

The report of the board in 1826 departed from Gallatin's early plan for a through route to and from the Tennessee River, favoring instead the development of the southeastern river systems of the Ocmulgee and the Oconee rivers as they joined to form the Altamaha River. Such improvements, they believed, would speed development in Georgia, supplement the steamboat river trade, and develop the seaports, which were considered more important than a pathway to the West. The board thought internal improvements should be made with state assistance to local projects rather than state construction of canals. Milton Heath has noted a paradoxical result. The board's opposition to the proposals to join the Tennessee River and the Atlantic Ocean by canal at that time and its partial approval of the railroad undoubtedly killed the greater canal movement in Georgia and possibly saved the state from a debilitating waste of money and effort.[62] Heath's judgment is reminiscent of that of Edward C. Kirkland, who noted the delay in these same years in Massachusetts when Boston hesitated to build a canal to link up with the Erie Canal at Albany.

Only two canals of significance were constructed in Georgia. One was the essentially coastal Savannah, Ogeechee and Altamaha Canal. On the Ogeechee trade floated to Ossabaw Sound, where it met coastal vessels, and a fifteen-mile canal linked the Ogeechee and the Savannah rivers. Construction began in 1825, with the work done largely by Irish laborers, and after five years the canal was finished at a cost of only $175,000. It was forty-eight feet wide and five feet deep with three wooden lift locks. Its chief engineer was Irish-born Edward Hall Gill. As a young man he worked with Benjamin Wright on the Erie Canal and then went on to work on the Morris Canal in New Jersey. After his service on this short Georgia canal, he returned north to work on canals in Pennsylvania and Ohio.[63] Work on the larger project to reach the Altamaha River, estimated to cost $625,000, continued through the 1830s. But before that destination could be reached, railroad competition caught up with the canal and the project was abandoned.

Georgia's second canal, the Brunswick Canal, never opened. A company chartered in 1834 and dominated by Boston investors sought to take advantage of the superior deep-water harbor at Brunswick, only twelve miles south of Savannah, and to garner the trade of the Altamaha basin before it could be tapped by Savannah. The company brought Loammi Baldwin, Jr., in 1836 from his work on the dry docks at Norfolk. He located the twelve-mile route of the canal to join "the Altamaha River, which had no good harbor, with the harbor of Brunswick which had no navigable river."[64]

Baldwin shared the optimism of the Georgians for the potential of water transportation to the interior. He reported that "the Altamaha River is navigable from Darien two hundred miles to the forks of the Ocmulgee and the Oconee, and up the Ocmulgee three hundred miles to Macon, and the Oconee two hundred miles to Milledgeville, the capital of the state." A trade of 130,000 bales of cotton, which he noted reached Darien from the rivers in the previous year, could be brought to Brunswick harbor. His canal had generous dimensions, fifty-four feet wide and six feet deep, with a lock at each end. But Baldwin fell ill before the canal could be begun, and he died in 1837. The company dug the canal, starting with hired slave labor and then turning to the Irish workers employed in 1839, who finished the job. But it was labor wasted for the emergence of railroad competition caused the canal to go unused.

A substantial canal project was also attempted in the bayou country of south-central Louisiana. State subsidies to the Barataria and Lafourche Canal Company, chartered in 1829, reflected national patterns of governmental assistance to canals. The plan was to dig a series of canals which would make a navigable route for some seventy miles from the Mississippi River to Bayou Lafourche. In spite of investments by prominent landowners, generous state assistance, and the labor of hundreds of slaves, the canal company could not complete the waterway. Construction that began in the 1830s created a limited navigation on parts of the route, but the state abandoned support in 1858. Robert Ruffin Barrow became sole owner and the Barrow family struggled to complete and improve the canal over a period extending well into the twentieth century.[65]

The canals from the Chesapeake to Georgia demonstrated an enthusiasm for canal technology almost equal to that in the North. They included the great national waterways of the Chesapeake and Ohio and Chesapeake and Delaware canals and pioneer waterways such as the James River and Santee canals. Precursors to the Intercoastal Waterway of today, the North Carolina canals were part of a north-south system from the Carolina sounds to the Chesapeake and across New Jersey to New York. Men such as Joseph C. Cabell of Virginia made canals a lifelong concern. Public figures prominent in other realms such as Charles F. Mercer, John Marshall, and James Monroe sponsored canals. The South Atlantic states had the same hopes as those of the mid-Atlantic coast for canals that would draw trans-Appalachian trade to their ports. Northern canal engineers such as Benjamin Wright and Loammi Baldwin brought their engineering expertise south, and southern states also drew on English canal experience.

But there were differences, too. Slave labor was often used to build southern canals. Although southern states were not averse to the assistance of state or national governments for canal building, most canals were the work of private companies, and most of them left balance sheets without profit. The southern canals operated year-round, but in general their traffic did not match that of canals farther north. Southern canals shifted sooner from an emphasis on long-distance routes to local development. The immigrant travel, so conspicuous on northern and western canals, was absent in the South. The states of the Old Northwest, however, would attempt the most ambitious expansion of the American canal system.

5

Canals of the Old Northwest

Most of the canals of the Old Northwest joined the Ohio River and the Great Lakes, and together they formed a great regional network. Yet they were also extensions of the canals in eastern states. Canal advocates in Boston, New York, Philadelphia, Baltimore, and even in the port cities of the South Atlantic hoped to capture the trade of Lake Erie and the Ohio Valley. The Erie Canal reached to Buffalo on Lake Erie, and the Pennsylvania Mainline system ended at Pittsburgh on the Ohio River. Steamboats appeared on Lake Erie after 1816, and the upstream steamboats on the Ohio after 1815 made it almost inevitable that the states of the Old Northwest would seize with enthusiasm on the promises of canals.

The result was the long canals between the Ohio River and the Great Lakes, with additional laterals and feeders. But almost equally important, short canals could surmount the old portages that had long impeded western travel. The portages at the falls of the Ohio, St. Mary's Falls in Michigan, and on the Des Plaines River near Chicago would all be passed by canals.

The Louisville and Portland Canal around the falls of the Ohio was built on the Kentucky side of the Ohio River and therefore not strictly part of the Old Northwest. An Indiana company was chartered for the purpose of bypassing the falls as early as 1805, and an abortive construction effort began on the Indiana side in 1819. The two-and-one-half-mile canal that was finally finished on the Kentucky side of the river in 1831 was vital to the steamboat traffic of the Old Northwest. Cincinnatians were deeply involved in this Kentucky venture because few upriver steamboats could pass beyond Louisville before the canal was built. After it was completed, upriver steamboats began to arrive in Pittsburgh from New Orleans in great numbers.[1]

After a successful challenge by Henry Shreve, Robert Fulton, Robert Livingston, and their heirs lost their claims on steam navigation of the western rivers. With steamboat operation open to all, tonnage increased dramatically. The cumulative effects of rising steamboat traffic on the

Ohio River, the appearance of the steamboat on Lake Erie, and the dramatic progress of the Erie Canal toward its western terminus at Buffalo gave an irresistible impulse to canal building in Ohio.

In 1818 Governor Ethan Allen Brown urged the Ohio legislature to begin surveys for canals from Lake Erie to the Ohio River. But any such canals would have to surmount the high ground that stretched across the middle of the state. In the Northeast, along today's U.S. Route 82, the same ancient rim of Lake Erie which the Erie Canal had vaulted at Lockport in New York would have to be surmounted by any canal leaving Lake Erie. As Frank Wilcox has described vividly, still higher and more hilly ground lay between the present locations of Akron and Columbus. And in the western part of the state, high ground near Salina marked the divide from which the Miami River ran south to the Ohio Valley and the Maumee River ran north to Lake Erie.[2]

The Ohio survey law of 1822 was the work of a committee headed by Micajah Williams of Cincinnati, a city dependent on the unpredictable New Orleans market yet opposed to a canal route through eastern Ohio. Williams's committee proposed five possible routes between the Ohio River and Lake Erie. Three New York engineers, James Geddes, David S. Bates, and William Price, began the surveys, and Alfred Kelley of Cleveland and Micajah T. Williams became acting commissioners. Kelley and Williams made a unique team, and their remarkable service to the state soon made them known as the fathers of the Ohio canal system. They went to New York and inspected the Erie Canal as a model, made surveys themselves, shrewdly compromised urban and regional rivalries, and secured legislative approval for two canals in 1825.[3] One would take an eastern and central path, and the other would be built in the southwestern corner of Ohio, close to the Indiana border. De Witt Clinton came in 1825 to break ground, first south of Newark for the Ohio and Erie Canal and then at Middletown for the Miami Canal.

The Ohio and Erie Canal combined parts of two of the originally proposed routes: the northern portion of a Cuyahoga-Muskingum line, and the Scioto Valley to the south, which had been part of a more central Scioto-Sandusky route. Beginning at Cleveland, the Ohio and Erie Canal ran up the Cuyahoga Valley to the Portage Summit, lifted by forty-two locks in thirty-eight miles to reach the summit at Akron, 309 feet above Lake Erie. Then the canal followed the Tuscarawas Valley south to Roscoe, where the Tuscarawas joins the Walhonding River to form the Muskingum River. The route ran along the Muskingum to Dresden but

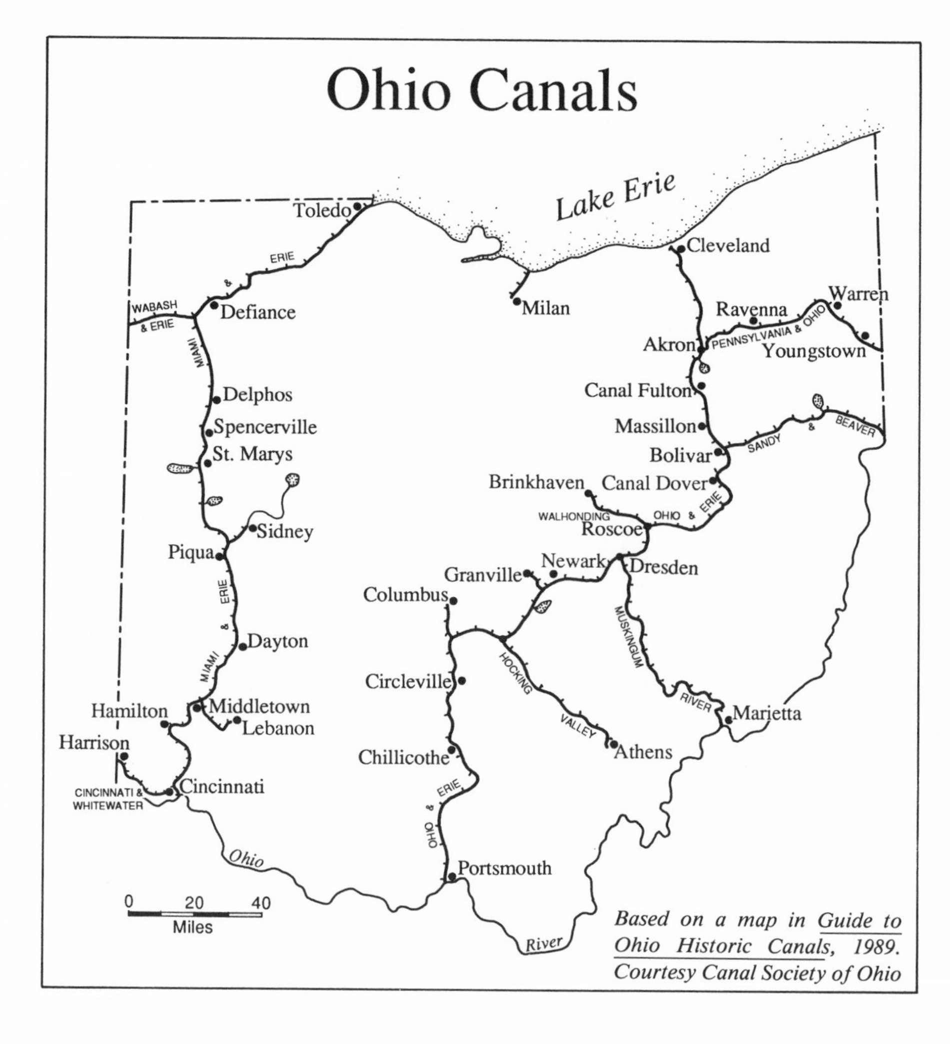

Based on a map in Guide to Ohio Historic Canals, 1989. Courtesy Canal Society of Ohio

did not continue south the short distance to Marietta as might have been expected. Instead, it turned west toward the center of the state and ascended the valley of the Licking River. West of Newark it crossed over the Licking summit to the Scioto Valley and, with a navigable feeder from Lockbourne to Columbus, followed the Scioto Valley almost due south to Circleville, Chillicothe, and, finally, Portsmouth on the Ohio.

The line between the Muskingum and the Scioto was most difficult to build. At the narrows of the Licking Valley, where the river flowed through a rocky gorge, the river was dammed and a towpath was blasted out of a sheer rock wall. Along the Licking summit there were three aqueducts on the fourteen-mile level. At the ridge between the Muskingum and Scioto rivers, the Granville feeder and the Licking Reservoir (now known as Buckeye Lake) supplied the Licking summit, and south of Millersport a three-mile-long deep cut was needed for water to flow from the reservoir into the canal.

Alfred Kelley was acting commissioner on the northern section of the Ohio and Erie Canal, and Micajah T. Williams directed work on the southern section. A group of Ohio engineers emerged to take their places alongside the engineers from New York, among them Francis Cleveland, Jesse L. Williams, and Richard Howe. Howe held the position of resident engineer on the northern division for twenty-one years after 1829. Jesse L. Williams made the final location of the line on the Licking summit and superintended construction in the Scioto Valley. The completed canal from Cleveland to Portsmouth was 308 miles long, had 151 locks, and passed over fourteen aqueducts. Its $4.3 million cost was less than that of any other canal of like mileage in America.[4]

The Miami Canal in the western part of the state began at Cincinnati, entered the Mill Creek Valley, continued northward, and passed through Lockland, Lesourdsville, Middletown, and Franklin to reach its terminus at Dayton. The first surveys were done by the New York engineers David S. Bates and Nathan S. Roberts. Micajah T. Williams was acting commissioner on the Miami Canal, and Samuel Forrer was the engineer in charge. Byron Kilbourne and Jesse L. Williams were assistant engineers. The sixty-six-mile canal with its fifty-four locks and five aqueducts was completed in 1828 at a cost of $900,000. Later it would be extended to meet the Wabash and Erie Canal near Defiance, which had its terminus at Toledo on Lake Erie.[5]

These canals were patterned on the Erie Canal, forty feet wide and four feet deep, with locks ninety feet long and fifteen feet wide. Ohio also copied the New York practice of having canal commissioners to supervise its canals and financing them by having commissioners of a canal fund sell bonds to banking houses in Ohio, New York, and abroad.

These first canals in Ohio were the result of mercantilistic planning for development of the state through public works. But soon after they were completed, egalitarian pressures and a belief that all should share in the benefits of public enterprise led to an extended program for lateral canals tied to the more viable trunk-line canals. New canals were approved in 1836, and the Loan Law of 1837 allowed the loan of public funds to private companies in a new policy of mixed enterprise.[6] At the same time, the legislature replaced the canal commission that had served since 1825 with a six-member Board of Public Works to manage the internal improvements of the state.

The costly Miami Extension Canal was opened in 1845, aided by a precedent-setting land grant from Congress in 1827. It ran from the Miami Canal terminus at Dayton past the Piqua farm of John Johnston, the canal commissioner, to the Loramie summit. Then the route followed the Auglaize and Maumee valleys to join the Wabash and Erie Canal with its terminus at Toledo. The distance on the Miami Extension Canal was 160 miles, and 103 locks were needed to overcome the rise and fall of 890 feet, making construction difficult.

David S. Bates and James Geddes made the initial surveys, and the engineers in charge were Samuel Forrer, William H. Price, and Jesse L. Williams. Once the challenge of a railroad alternative along the Mad River was disposed of, the problem of the high ground at the divide, the watershed between Lake Erie and the Ohio River, had to be overcome. To supply water to fill the canal in both directions, the engineers turned to the construction of a series of great reservoirs.

After crossing Loramie Creek, going north, the canal approached the summit rising on a flight of five stair-step locks at Lockington. The Sidney feeder brought water from the Miami River and entered the canal at the south end of the summit. The town of Port Jefferson, near Sidney, grew at the highest point on the feeder, attracting a grain and lumber trade that could be readily carried on its fifty-foot width. Construction was delayed, however, on the great reservoir farther up the Miami River above Sidney, which was to supply water to the Miami Extension Canal through the feeder. It was not finished until 1852, but it ultimately covered more than six thousand acres.

At the northern end of the Loramie summit, the Loramie Reservoir added water to the canal in 1843, and after the canal dropped through eleven locks in six miles, the Grand Lake St. Mary's Reservoir supplied it with water north to the junction with the Wabash and Erie Canal. These great works contributed to the high cost of the Miami Extension, $3,195,000, a million dollars over the estimate and more than any other canal of equal length in Ohio.[7]

While canal construction was under way in northwestern Ohio, in the extreme southwestern corner of the state the Cincinnati and Whitewater Canal was being built by a private company and was completed in 1843. This twenty-five-mile canal was surveyed by Darius Lapham, who was also a resident engineer on the Miami Canal. West of Cincinnati it crossed the Miami River on an aqueduct and then used a daring tunnel that was dug sixteen hundred feet through a ridge. The canal passed in front of the home of William Henry Harrison and connected to the Whitewater Canal in Indiana. The Cincinnati and Whitewater Canal was owned chiefly by Cincinnatians, and it made Cincinnati (after Lawrenceburg) a second terminus of Indiana's Whitewater Canal.

From Middletown, where construction on the Miami Canal had begun, the short-lived (1840-48) Warren County Canal ran seventeen miles east to Lebanon on the old stage route to Cincinnati. It reached into a rich agricultural region and passed by the now vanished Shaker settlement in Union village. It cost $217,600, and it was perhaps the least successful segment of the Ohio canal system.[8]

Up the Ohio River, in southeastern and eastern Ohio, there were three lateral canals along with the canalization of the Muskingum River, which created new connections to the Ohio and Erie Canal. The Muskingum River was canalized from Marietta to Dresden, and an additional canal was built from Dresden Junction to the Ohio and Erie Canal to complete a ninety-one-mile waterway with twelve locks that was finished in 1841. The through boats arriving at Marietta from Cleveland the following spring thus took a far shorter route than if they had stayed on the Ohio and Erie Canal and followed the Scioto Valley to Portsmouth. Moreover, the Muskingum improvement took steamboats, which made daily runs from Marietta to Dresden, some built especially to tow boats.[9]

In 1843 the Hocking Valley Canal was added, leaving the Ohio and Erie Canal at Carroll and continuing fifty-six miles, with twenty-six locks, passing through the coal- and salt-mining region to Athens. North of these southeasterly laterals, the Walhonding Canal ran for twenty-five miles northwest from Roscoe, up the Walhonding Valley toward the agricultural center of the state, opened in 1841 and using eleven locks.

The shortest routes to the Ohio River from the Ohio and Erie Canal were in the northeastern corner of the state. Ohio built two more canals that could link up with the Ohio River and with the Pennsylvania canals as well. The Sandy and Beaver required only seventy-three miles of canal between the Ohio and Erie at Bolivar and East Liverpool on the Ohio, only forty miles below Pittsburgh. It was begun in 1834 by a private company aided by the state. But the company ran out of funds during the depression of 1837 and suspended operations. When work was resumed in 1845, a difficult tunnel construction delayed completion until 1850, and it was in full service for only two years. The Sandy and Beaver Canal attracted little traffic partly because farther north the Pennsylvania and Ohio Canal was completed in 1840, running almost parallel on a route to the Ohio River. Often called the Mahoning Canal, it was also built by a private company aided by the state. The section leaving the Ohio and Erie Canal at Akron passed through Warren and Youngstown to the Ohio line and was completed in 1839. The Pennsylvania portion, from the Ohio line direct to Pittsburgh, was finished the next year.

This canal linked with the Pennsylvania Mainline and intersected with the Erie Extension Canal, which terminated at Lake Erie. Thus the Erie Extension Canal and the Ohio and Erie Canal ran parallel to each other, on either side of the state line only a few miles apart, just as in the southwestern corner of Ohio, the Miami and Whitewater canals ran parallel a few miles apart on either side of the Ohio-Indiana line. In northeastern Ohio, the interlaced cross-border canal network led to a highly competitive contest for the cheapest routes to Lake Erie or to the Ohio River.[10]

At the lake port of Huron in northern Ohio, the Milan Canal extended only three miles into the wheat-growing region and had only two locks. But when it was opened in 1840, its mile-long ship canal took the largest lake vessels and helped make Milan a great interior wheat port. Exports of wheat and flour rose from 4,000 bushels to a peak of 1.6 million bushels in 1847.[11]

The Ohio and Erie Canal carried more trade than any other Ohio canal. In 1851 2,529,342 bushels of wheat arrived at Cleveland, 645,730 barrels of flour, almost a million bushels of corn, almost 3 million bushels of coal, and nearly 11 million pounds of merchandise was shipped southward. Moving south in 1851, canal trade arriving at Portsmouth included almost half a million bushels of corn, more than 70,000 bushels of wheat, almost 50,000 barrels of flour, and almost 5 million pounds of pork and bacon, while 2.6 million pounds of merchandise moved north up the Ohio and Erie Canal.[12] One can imagine the numbers of boats, each laden with 50 to 80 tons of goods, moving in silent procession at four miles an hour, locking through to higher or lower levels, to carry so vast a trade.

More grain came to Cleveland from the interior of the state than from the Cuyahoga Valley or the Lake Erie shore. Wheat and flour came from the Scioto, Muskingum, and Tuscarawas river valleys. John G. Clark has calculated that the wheat and flour exports from Newark, Dresden, Roscoe, Dover, Massillon, and Akron constituted almost all of the receipts at Cleveland. Akron milled much of its wheat into flour and sent 376,000 barrels of flour to Cleveland in 1846 and 1847. In 1847, Massillon sent 1.7 million bushels of wheat and flour north on the canal. Dover in the Tuscarawas Valley sent approximately 54,000 barrels of flour and 410,000 bushels of wheat to Cleveland each year between 1843 and 1851.[13]

The Miami and Erie Canal was less heavily used than was the Ohio and Erie, but its trade was a major impetus to the rise of Cincinnati as the Queen City of the West. It carried more flour, pork, and whiskey than bulk wheat. In fourteen of the seventeen years from 1833 to 1850, between 18,000 and 35,000 barrels of pork came to Cincinnati on the Miami Canal. In 1851 the canal brought to Cincinnati 130,292 bushels of wheat, 317,107 barrels of flour, and 270,147 bushels of corn. Dayton served as terminus of the Miami Canal before the extension to Toledo was completed and furnished half of the flour shipped to Cincinnati. A sense of this busy trade is reflected in the traffic recorded for the first month of the season of 1830, when seventy boats departed Dayton and seventy-one arrived. Going north, 10 million pounds of merchandise cleared Cincinnati on the canal in 1846 and more than 6 million pounds in 1851. When the Miami Extension was added, Cincinnati sent southern sugar and molasses to Toledo and to Indiana, a million pounds going north in 1848.[14]

Near the middle of the Miami Canal, between Cincinnati and Dayton, Hamilton on the Miami River was connected to the canal by a mile-long basin. Daniel Preston has traced the changes brought about by the canal in that Miami Valley village.[15] Before the canal was completed to Hamilton in 1828, its river trade found a market by flatboats to New Orleans, with little going to Cincinnati, which was several miles up the Ohio River from the mouth of the Miami. By 1833, with the canal in operation, daily freight boats ran south to Cincinnati, which replaced New Orleans as Hamilton's primary market. Even after the Miami Extension Canal was opened in the late 1840s, the pattern of trade to Cincinnati did not change. Some goods were sent to Lake Erie, but only in small quantities. Instead, after the completion of the Pennsylvania canal network Hamilton's trade to the east went up the Ohio and across Pennsylvania rather than north and east by the Lake Erie route.

Tonnage figures show Hamilton's substantial contribution to Miami Canal traffic, rising from 520 tons in 1833 to almost 3,400 tons by 1860. More than 4,000 tons of goods arrived at Hamilton by canal in 1851. Hamilton became a packing center, sending pork to the larger packing center in Cincinnati, the 6 million pounds in 1841 representing about 32,000 hogs. More than 2 million pounds of lard were shipped by canal in 1844. Flour was less significant, with a high figure of almost 65,000 barrels in 1847, and corn exports by canal reached almost 144,000 bushels in 1853. The population of Hamilton quadrupled, from 1,708 in 1830 to 7,500 in 1860.[16]

Most of the lateral canals in Ohio were financial failures. In southwestern Ohio, the Warren County Canal was abandoned after only eight years of use. In eastern Ohio, the Hocking Valley and the Walhonding canals never made an appreciable return on their investment; nor did the Pennsylvania and Ohio and the Sandy and Beaver canals achieve the levels of traffic anticipated. The latter involved the cities of northeastern Ohio in a commerce as delicately balanced as that at the head of Delaware and Chesapeake bays. Depending on tolls, products, and state policy, goods on the Ohio and Erie Canal had the options of reaching the Ohio River or Lake Erie by the Ohio lateral canals or by the Pennsylvania laterals, and they had alternate routes to the east by the Erie Canal or the Pennsylvania Mainline.[17] Traffic on all the Ohio canals peaked in 1851, and the railroad competition brought about a decline. Ohio might appear to have been the preeminent canal state west of the Appalachians. Altogether, the state of Ohio constructed more than eight hundred miles of canals at a cost of almost $16 million.[18] They have attracted detailed study of their land grants, profit and loss, and relation to government.

Yet it was in Indiana that the new canal technology was applied most dramatically to frontier conditions. Indiana became a state the year before construction started on New York's Erie Canal and only six years before Ohio authorized its canals. In this new frontier state, plans for canals were cast in a grandiose design that would penetrate far into the wilderness, projecting a canal line reaching from Lake Erie to a point far down the Ohio River Valley, only a few miles from the Illinois border. These canals were the work of a speculative legislature, supported by vast national land grants and carried out under a scarcely believable policy that allowed all of the canals to be worked on at once.

The historic Wabash trade route, which ran southwest from the Maumee River in Ohio, provided the natural path for the longest single canal in the Canal Era, the 468-mile-long Wabash and Erie Canal. The ease of connecting the Maumee and the Little Wabash rivers over the seven- or eight-mile portage near Fort Wayne made such a canal almost foreordained. Learning of the plan, De Witt Clinton wrote to Benjamin F. Stickney in Indiana, "I have found the way to get into Lake Erie and you have shown me how to get out of it. . . . You have extended my project six hundred miles."[19] The Wabash and Erie Canal began with an Ohio portion at Toledo, crossed the summit near Fort Wayne, and continued on to Evansville on the Ohio River. When construction began in 1832 near Fort Wayne, most of Indiana's population was in the southeastern corner of the state. Though Ohio delayed work on its Maumee section, the canal reached Peru in 1837, Lafayette in 1843, Terre Haute in 1848, and, at last, Evansville, in 1853.

In Indiana, the pioneer entrepreneur who threw his energies into the Wabash Canal project and saw it through to realization was Samuel Hanna. His canal career resembled those of Alfred Kelley and Micajah T. Williams in Ohio and other men who worked on canals in the East. Hanna arrived in Fort Wayne as a fur trader in 1819. Along with John Tipton, the Indian agent and land speculator at Fort Wayne, Hanna led in the canal movement. He became a canal commissioner, did his own surveys, served as a Canal Fund commissioner, and assisted in organizing the State Bank of Indiana. He speculated in land around Fort Wayne and far down the Wabash Valley; he owned stores and held milling interests in Fort Wayne, South Bend, Lafayette, and Wabash; and he moved on to railroads when they began to supplement or replace canals.[20]

But more than the personal advocacy of Samuel Hanna, canal building in Indiana was fueled by federal land grants. In 1827, when a grant was given to aid the Illinois and Michigan Canal, a donation was

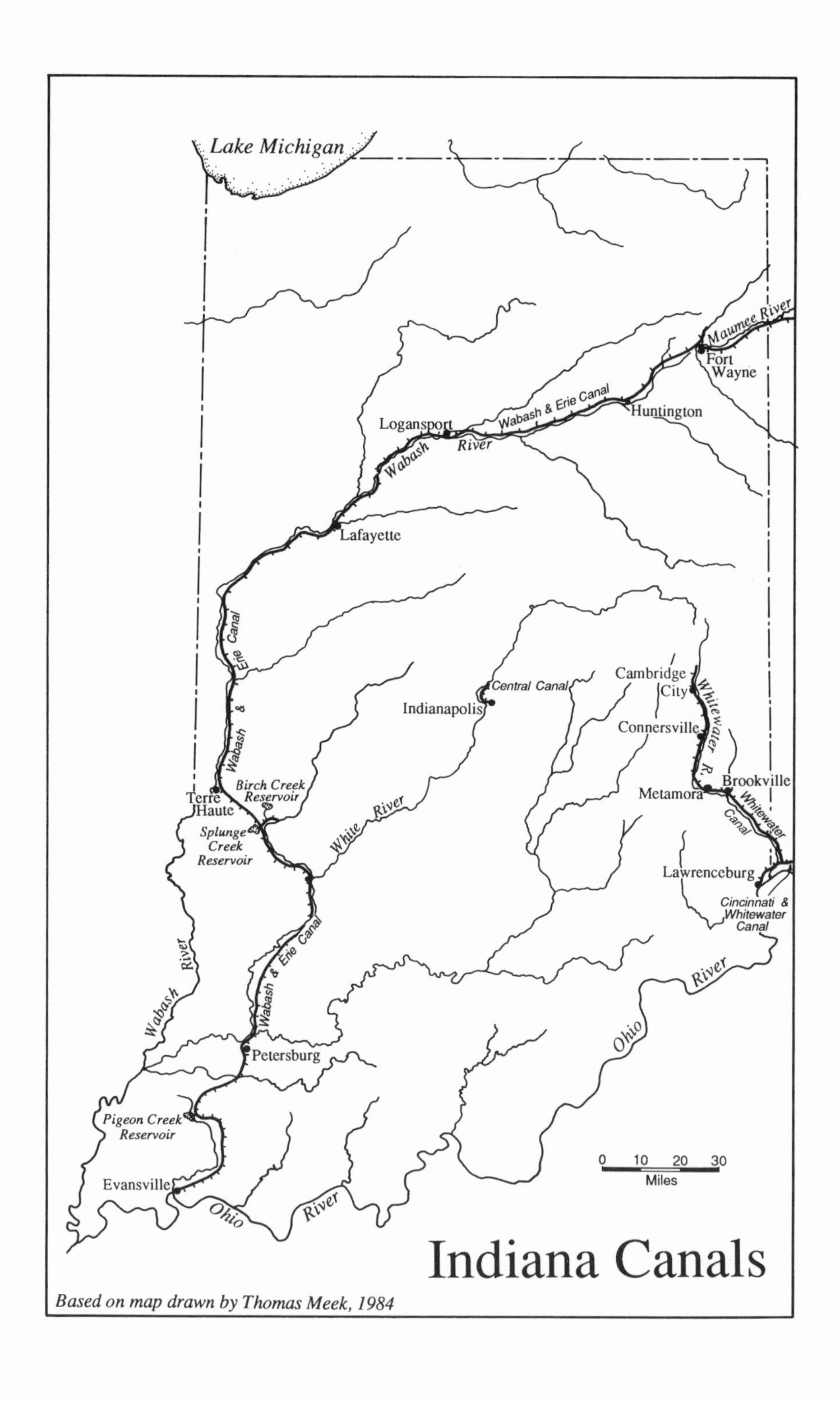

Lake Michigan
Maumee River
Fort Wayne
Wabash & Erie Canal
Huntington
Logansport
Wabash
River
Lafayette
Wabash & Erie Canal
Central Canal
Indianapolis
Cambridge City
Whitewater R.
Connersville
Brookville
Metamora
Whitewater Canal
Lawrenceburg
Cincinnati & Whitewater Canal
Terre Haute
Birch Creek Reservoir
Splunge Creek Reservoir
White River
Wabash River
Wabash & Erie Canal
Petersburg
Pigeon Creek Reservoir
Evansville
Ohio River
Ohio River
0 10 20 30
Miles
Indiana Canals
Based on map drawn by Thomas Meek, 1984

given to Indiana as well. The Indiana grant gave five sections of land, totaling thirty-two hundred acres, for each mile of canal constructed; the land was in alternate sections on both sides of the canal in a checkerboard pattern similar to the railroad grants later in the century. Acceptance of the grant required the building of a canal within five years, and this stipulation provided the motive, even the obligation, for the great expansion of the Indiana canal system.[21]

Since fulfillment of the terms of the land grant required that the canal pass through Ohio to Lake Erie, Ohio agreed (in return for part of the grant) to build the section from the Indiana border to Junction, Ohio, just south of Defiance, where the Wabash and Erie Canal joined the Miami and Erie Canal to reach Lake Erie. Yet in a perverse fear of potential competition from the Indiana canals, Ohio did not complete her eighty-eight-mile portion until 1843.

Moral or legal obligations to construct a canal with the land grant of 1827 aside, Indiana confronted decisions similar to those faced in Philadelphia, Baltimore, Boston, and elsewhere: whether to build a canal or try the unproven technology of the railroad. Personally favoring railroads over canals, Governor James B. Ray included the railroad option in his message to the Indiana legislature in 1827 and estimated that it would be only half as costly to build a railroad as a canal.[22] But the "Wabash Band" in southwestern Indiana vociferously supported a canal, and the railroad interests were outnumbered. Moreover, it was argued that the canal could be made with local materials such as the famous Indiana limestone, whereas tracks and railroad cars must be imported. The more conservative choice of a canal won out. As Ralph D. Gray has written, "At no time, in fact, was any serious consideration given to substituting a railroad for any portion of the Wabash and Erie Canal."[23]

Moreover, national canal enthusiasm peaked in the mid-1830s with the Loan Law expansion of Ohio canals in 1836, the enlargement of the Erie Canal in New York, and the mountain crossing of the Pennsylvania canals. It is little wonder that Governor Noah Noble urged the legislature "to follow the successful examples of other states" and that in 1836 the legislators passed a bill providing for a "general system of internal improvements." Soon known as the Mammoth Internal Improvement Act, the measure supported eight turnpike, canal, and railroad projects at an estimated cost of $10 million. If this was overly ambitious, more disquieting was the provision that work should begin on all projects at the same time.[24]

More than one-third of the $10 million went to the Wabash and Erie Canal, to build as far as Terre Haute. The allocation reflected broad support for this project even though it was to be built in the less settled northeastern part of Indiana. But two other canals in the "general system" resulted from regional rivalry, the Whitewater and the Central canals, linked to each other and to the Wabash and Erie Canal. The former was located in the more settled southeastern corner of the state, from Lawrenceburg on the Ohio River (only a few miles from the Ohio line) north through Brookville, Metamora, Laurel, and Connersville to Cambridge City on the National Road. This canal climbed a formidable 491 feet in only seventy-three miles.[25] In much the same way that Ohio had built the Miami Canal in a sectional trade-off with the Ohio and Erie Canal, the Whitewater Canal was begun as a trade-off with the Wabash route; but it was also a constant competitor with the Wabash and Erie Canal for state funds.

Indiana's ill-fated Central Canal was designed to supplement the Wabash and Erie, though only a small portion along the White River through Indianapolis was ever fully completed, some of which may be seen today, where it serves for water supply and recreation. Leaving the Wabash and Erie near Peru, it was to make a great easterly loop to Indianapolis, pass on through Martinsville and Spencer to Worthington, and then continue through Petersburg to Evansville on the Ohio River. A cross-cut canal from Terre Haute was to link the Wabash and Erie to the Central Canal from Terre Haute to Worthington, but this and the Central Canal route below Worthington ultimately became part of the Wabash and Erie Canal. In this grand scheme a never constructed eastern branch of the Central Canal from Anderson to North Muncy would have linked with the Whitewater Canal.

The Mammoth Bill in Indiana created a Board of Improvement and a Board of Fund Commissioners, which resembled the administrative structure in Ohio. But where Ohio had entered its canal program with remarkable dispatch, Indiana started its program at sixes and sevens. There was immediate rivalry among the acting commissioners for funds, the fund commissioners borrowed money in New York at excessive rates, and a general laxity pervaded disbursements and record keeping.[26]

The principal engineer was Jesse L. Williams, who had learned from Samuel Forrer in Ohio. Appointed in 1832, he moved to Fort Wayne and surveyed both the Wabash and Erie Canal and the Whitewater Canal. In

1836 he was named chief engineer of the state of Indiana and later became a member of the Board of Improvement.

With five thousand men at work, the Wabash and Erie Canal was completed from Fort Wayne to Lagro in 1837 and on to Logansport in 1838. The familiar original Erie Canal prism was adopted, forty feet wide and four feet deep. Locks measured ninety by fifteen feet. Along the Maumee and Wabash river valleys, the route was low and flat. Yet this long-distance canal crossed innumerable streams and is noted for its culverts. Unimposing as they may appear, there were 144 culverts in the first 160 miles. Indiana was blessed with abundant high-quality limestone, yet canal stonework came from whatever quarry was near the canal structures, and the results were varied. The Georgetown quarry below Logansport, for example, yielded excellent stone and produced lasting structures, but the Salania quarry above Lagro produced stone that fractured under pressure in canal structures and often required rebuilding. Some aqueducts such as the Aboite aqueduct eleven miles southwest of Fort Wayne were built on piers and abutments made of timber because of the poor quality of stone available in parts of the Wabash Valley.[27]

Meanwhile, work was started on the Central Canal, and by 1839 the Whitewater Canal was finished up the steep valley of the Whitewater River from Lawrenceburg to Brookville. Before all work on the Indiana canals was suspended in 1839, the expenditures on Mammoth Bill projects had reached almost $8 million for about 140 miles of canals and 70 miles of railroads and turnpikes.[28]

The collapse of canal construction in Indiana was a result of the Panic of 1837 and the failure of the Morris Canal and Banking Company in New Jersey. Indiana sold bonds to that banking company on credit terms bordering on fraudulence, or worse, and was therefore especially vulnerable to the depressed economy of the times. The state lost $3.5 million from the failure of the Morris company and other creditors. By 1841 Indiana was insolvent, with a debt of $13 million, $9.5 million of which had been spent on canals and railroads.[29] Abandoned canals were sold to private companies.

Only the prized Wabash and Erie Canal went on haltingly, its workmen paid by canal scrip known as "White Dog" or "Blue Dog" instead of cash, so as to carry out the obligations of the canal land grant.[30] By 1841 navigation was opened from the Ohio-Indiana line to the Tippecanoe River, reaching Lafayette in 1843. Another congressional land grant of 260,000 acres promised construction west to Terre Haute; and in the east, the long-delayed completion of the Ohio portion of the Wabash and Erie Canal to Toledo opened 175 miles of canal from Indiana to Lake Erie. Still a third congressional land grant came in 1843, influenced in part by the twenty miles on the southern division of the Central Canal completed and in operation between Evansville and the Pigeon Creek Reservoir in Gibson County. This time Congress gave alternate sections on the canal line and all the unsold land in the Vincennes Land District, which enabled construction that would carry the canal from Terre Haute to Evansville on the Ohio River. Construction reached Terre Haute in 1849, and the great diagonal canal line crossing the state from Lake Erie to the Ohio River was completed by 1853.

Crucial to this achievement were the efforts of Charles Butler of New York, who came to Indiana and sought to recover some of the losses of British bondholders. In the legislature the "Butler bills" of 1846 and 1847 put the Wabash and Erie Canal into the hands of a Board of Trustees who would operate the canal and extend it to Evansville. The trustees added $2,375,236 to the $6,437,809 already spent by the state. They were little short of heroic in their efforts. They faced the cholera epidemic on the canal line in 1849-50, which took the lives of 150 laborers and one of the trustees, delayed construction, and depressed revenues. Butler's belief in the special advantages of canals sustained his work much as did that of Joseph Cabell in Virginia on the James River and Kanawha Canal. Management of the canal by the trustees was exemplary, and traffic on the Wabash and Erie reached its peak in 1852. The trustees struggled on, watching railroads parallel almost the entire canal line in the 1850s. Butler's management continued until after the Civil War. Meanwhile, the Wabash and Erie Canal was leased in sections to private contractors who operated it until it was sold in 1874.[31]

In the period of trustee management, the completion of the 111-mile section of the Wabash and Erie Canal from Point Commerce below Terre Haute to Evansville was a notable achievement. It was decisive in gaining the last and largest land grant from Congress, yet it was the most troubled part of the canal line. The terrain was difficult, and two reservoirs were required to supply water, the four-thousand-acre Splunge Creek Reservoir on the Eel River summit level and the thousand-acre Birch Creek Reservoir. Instead of the local support typical of the Canal Era, here there was prolonged opposition which led to violence and vandalism.

In Clay County opposition arose from the fear of malaria and cholera that were believed to come from the canal reservoirs, especially where uncleared timber rotted beneath the water as was the case in the great Splunge Creek Reservoir. The trustees pointed to the absence of such effects on the Ohio reservoirs, which were equally large, and a board of physicians reported no dangers. Thomas Dowling, the resident trustee at Terre Haute, was assured that "the Trustees will yet be considered as public benefactors by restoring health to a large tract of country heretofore subject to the endemic diseases prevalent in the whole wet and marshy region now submerged." But the Clay County vigilantes, or "Reservoir regulators" as they were called, numbering perhaps a hundred men, cut the Birch Creek Reservoir repeatedly in 1854 and drove off the workers. Canal navigation came to a halt. The trustees sought to clear the timber from the Birch Creek Reservoir, and the governor called out the militia to guard the works. But in 1855 the regulators drained the reservoir and two years later burned an aqueduct. By 1860 the canal south of Terre Haute was virtually abandoned.[32]

The Wabash and Erie Canal has often been considered one of the supreme follies of the "canal mania" of the 1830s. George Rogers Taylor has written in his classic history of the transportation revolution of nineteenth-century America that "no state became more disastrously involved in the general enthusiasm for the canal building than Indiana." The Wabash and Erie Canal has been criticized for its imprudent if not fraudulent financing, its unprofitable extension by the trustees from Terre Haute to Evansville, and its minimal toll receipts. Tolls never rose above a pittance, reaching a high of $193,400 in 1853, followed by a long decline. Indeed, the main income from the canal came from the sales of land from the three land grants given to Indiana. Sales amounted to $5,477,238 against a total cost, by the time the Wabash and Erie Canal was sold in 1874, of $8,259,244.[33]

Yet traffic on the Wabash and Erie Canal was heavier than its checkered history might suggest. Evansville became a major river port in the lower Ohio Valley at the southern terminus of the canal. Between 1853 and 1859 it received by canal more than 250,000 bushels of wheat; almost 4.5 million pounds of bacon, pork, and lard; 250,000 pounds of furniture and an equal weight of cordage; more than a million staves and shingles; and the largest item of trade, almost 4 million pounds of tobacco. Evansville became a distribution point for coffee, sugar, and molasses. Nearly 3 million pounds of molasses and more than 1 million pounds of sugar went up the canal from Evansville.[34]

Canal traffic arriving at Toledo on Lake Erie, which included that on the Miami Extension Canal, reached impressive levels in 1851. Freight arriving in Toledo by canal included 1,639,744 bushels of wheat, 242,677 barrels of flour, and 2,776,149 bushels of corn. Tonnage moving east on the Wabash and Erie Canal rose from 184,400 in 1850 to a high of 308,667 in 1856. As in the case of Cleveland farther east, the Wabash and Erie Canal allowed Toledo to tap the wheat and flour production of an interior region stretching as far west as Cass County, Indiana. Before 1845, Lafayette had been shipping to the New Orleans market, but that year 539,000 bushels of wheat and flour and 1,313 tons of pork and bacon went northeast by canal. Only four years later, 8,740,000 bushels of corn were added to 600,000 of wheat and flour going northeast, along with increased quantities of pork, lard, and bacon.[35]

Moreover, the Wabash and Erie Canal joined with the Miami Extension Canal to provide a water route between northern Indiana and Cincinnati, which was significant to Cincinnati's canal exports. A poignant event occurred during a journey on this canal route in 1846 when 325 Indians boarded canal boats at Fort Wayne to take the Wabash and Erie Canal to Defiance, the Miami Extension Canal to Dayton, and the Miami Canal to Cincinnati in their removal from their Wabash lands to Kansas.[36]

As with other canals, the developmental impact of the Wabash and Erie Canal outweighed the financial losses incurred by the state. Land sales were not only a stimulus to canal construction but a result of the canal's influence. Population in the canal counties grew from 12,000 when the first section of the Wabash and Erie Canal was opened in 1835, to 60,000 when the canal was completed from Lafayette to Toledo, and to 150,000 after the entire line was finished. In 1854 a canal mail route extended from northern Indiana to Evansville. New cities were created which flourished long after the canal was gone, such as Fort Wayne, Huntington, Wabash, Peru, Logansport, Delphi, Lafayette, Covington, and Attica. Other cities flourished only as long as the canal operated such as Lagro, Lewisburg, Georgetown, Carrolton, Americus, Lockport, and Pittsburgh.[37] In spite of all its defects, Ralph D. Gray has concluded that the Wabash and Erie Canal was "the single most important development in the political and economic life of Indiana during the mid-nineteenth century."[38]

The Whitewater Canal occupies a singular place in the history of

Indiana canals. Its importance for the more heavily populated settlements in southeastern Indiana led R. Carlyle Buley to call it "the key to the whole logrolling process" in the enactment of the general system in 1836.[39] Begun by the state, it was finished in 1846 by a company that included Cincinnati investors. Its fifty-six locks carried the canal up the steep rise of the Whitewater Valley seventy-six miles to Cambridge City with seven feeder dams. They fought a losing battle with the floods that came in 1847, 1848, 1850, and 1852, the last of which brought canal operations virtually to an end. Income from tolls never rose above $30,000 a year, yet when it was serviceable it carried the agricultural produce of eastern Indiana to the Ohio and served its region sporadically until 1865. Its Duck Creek aqueduct at Metamora has remained famous as a covered bridge aqueduct, and a surviving grist mill reminds us of the waterpower taken from midwestern canals. It attracted the Cincinnati builders of the Cincinnati and Whitewater Canal along the Ohio River, who drew the trade of the Whitewater Canal to Cincinnati, and as late as 1849 another company extended the Whitewater Canal north to Hagerstown.[40]

A year after Indiana enacted its Mammoth Bill, the state of Illinois embarked upon an almost equally ambitious program of internal improvements. In 1837, a Whig legislature (which included Abraham Lincoln) authorized a myriad of transportation projects. But 1837 was a year of financial panic, and the only survivor of the Illinois canal projects was the Illinois and Michigan Canal, which had been authorized in 1835.

The Illinois River route from Lake Michigan to the Mississippi shared many of the characteristics of the Wabash trade route in Indiana or, farther north, of the route to the Mississippi from Lake Superior by the Fox and Wisconsin rivers. The portages on the Des Plaines River near Chicago, at Portage in Wisconsin, and near Fort Wayne in Indiana were fulcrum points for trade awaiting improvements by canals. The Illinois and Michigan Canal was ninety-six miles long, passing from the Chicago River across the summit to the Des Plaines Valley, following that river to the rapids of the Illinois River at La Salle, below which the Illinois River could be navigated to the Mississippi.

Illinois created a Board of Canal Commissioners in 1823 and chartered a company for such a canal in 1825, only to recall the charter in 1826. Governor Ninian Edwards helped secure a three-hundred-thousand-acre federal land grant in 1827, which served as a stimulant to canal building in Indiana. Nine years of controversy followed during which railroad proponents were defeated and sectional divisions were overcome. In 1835, with Governor Joseph Duncan's strong support, the legislature finally authorized the Illinois and Michigan Canal. A new act in 1836 pledged the credit of the state to the canal and created a board of three canal commissioners to carry out the work.[41] Construction started in 1836 on a canal that would begin at Chicago and terminate at LaSalle near the mouth of the Little Vermillion River.

Unlike the canals of Ohio and Indiana, but nearer the dimensions of the enlarged Erie Canal just begun in New York, the Illinois and Michigan Canal was 60 feet wide and 6 feet deep with locks 110 feet by 18 feet. William Gooding, an engineer from the Ohio canals, first surveyed the route, but work did not proceed until Benjamin Wright came from New York and endorsed the plan. One of the canal commissioners in 1836 was Gurdon S. Hubbard, whose career was much like that of Samuel Hanna in Indiana. Hubbard was a fur trader, merchant, and ship owner in the Lake Michigan trade, who later went on to a business career in the meat-packing industry in Chicago.[42]

In spite of the Panic of 1837 and minimal sales of canal lands, the commissioners of the Canal Fund in Illinois borrowed money in New York and London, and construction began. Gooding supervised the work as chief engineer, and E. B. Talcott was assistant engineer. But few canal workers were available locally from the sparsely settled lands along the route. Catherine T. Tobin has described the recruitment of Irish labor for the canal, which serves as a case history of Irish canal labor in America. Because of the dearth of local laborers, contractors advertised in newspapers in New York, and in 1839 the Board of Canal Commissioners hired an agent to travel "to any place within the United States" to recruit up to a thousand laborers and send them to Chicago. The commissioners reported in 1837 that the work was being pursued "with great vigor" though construction was hampered by short supplies and contractors lacked sufficient capital "in so new and remote a country." Although the state assisted the contractors with funds, supplies, and the larger canal-building machines, funds fell short. Workers were paid in 1839 in canal scrip, which depreciated to sixty-five cents on the dollar in 1840 and only thirty cents in 1842.[43] Nonetheless, by 1842 about three-fourths of the work on the canal was completed.

Then the failure of the State Bank of Illinois plunged the state into a crisis over the canal similar to that which brought Indiana to insolvency in 1841, Ohio to near insolvency, and the stoppage of work on canals in New York, Pennsylvania, and Maryland. Illinois defaulted on its bonds, and work on the Illinois and Michigan Canal was suspended in 1843.

Five years of wrangling and stalemate, involving demands for railroads as well as canals, delayed completion of the canal until 1848. But the fundamental issue was the settlement of the $15 million Illinois state debt held by American and foreign bondholders. Charles Butler represented the foreign bondholders in a settlement of the Indiana debt which provided for the completion of the Wabash and Erie Canal, and his friend Arthur Bronson represented the bondholders and drafted a similar plan for Illinois. Bronson joined with William B. Ogden, a principal contractor on the canal, to propose a plan by which the canal would serve as a guarantee for the debt. The canal would be deeded to trustees, who would supervise its completion. The plan also called for a property tax to satisfy the American and foreign bondholders that their interest would be paid. For their part, the bondholders would make a new loan to complete the canal. The Illinois legislature adopted most of the plan, agreeing to deed the canal to trustees and negotiate a new loan for $1.6 million. But the legislators refused to authorize a tax. Governor Thomas Ford appointed Charles Oakley, a former fund commissioner, and Michael Ryan, chair of the legislative committee on canals, to negotiate the loan. They secured a small loan in New York, but their journey to Europe was unsuccessful. Only after a state tax was added in 1845 did the European bondholders grant their loan of $1.6 million. The canal was promptly deeded to their trustees and a new tax law in 1847 saw the canal through to completion the following year.[44]

When the trustees resumed work in 1845, they abandoned the earlier plan for a deep-cut canal. But the gathering of a labor force was once again a major problem. An agent went east to recruit, and notices were placed in newspapers in New York and Boston as the trustees sought to hire between one and two thousand workers at an average wage of $20 a month. Robert Stuart, secretary of the Board of Canal Trustees, wrote in 1846 that "there is a great scarcity of men," and work was delayed by the annual "sickly time" and the abandonment of contracts by contractors.[45]

As secretary of the Board of Canal Trustees, Stuart kept the trustees informed of progress. There were three trustees, William Swift, who lived in Washington; David Leavitt, president of the American Exchange Bank in New York, who served as treasurer; and Jacob Fry, the state trustee. But when Charles Oakley replaced Fry in 1847, the trustees came into open conflict with each other and with William Gooding, the chief engineer. Oakley's largely political grievances were turned against his fellow commissioners and Gooding. Their troubles were compounded when they all became enmeshed in the strike led by Daniel Lynch, who represented Irish workers on the canal.[46] Though the trustees gave much dedicated service, there was less harmony and cooperation among them than that which had distinguished the trustees who finished the Wabash and Erie Canal. Gooding remained chief engineer until the canal was completed. He then became secretary of the board, and in 1854 he became general superintendent of the canal.

Boats could now move southwest up the Chicago River and follow the canal through Lockport, Ottawa, and Utica, to La Salle on the Illinois River, with easy passage to the Mississippi from there. With its fifteen locks and four aqueducts, the canal had cost $6.4 million.[47] Abraham Lincoln, now in Congress, announced its completion to the House of Representatives.

The Illinois and Michigan Canal carried a trade that rose rapidly for more than a decade. From 1848 to 1850, wheat and corn moved in about equal amounts, but in the 1850s the canal facilitated the phenomenal increase in corn production in northern Illinois. In 1851 the canal carried to Chicago 2,878,550 bushels of corn, 78,062 bushels of wheat, 4,591,471 pounds of sugar, and 56,845,027 feet of lumber; to the interior it carried 14,175,928 pounds of merchandise.[48] The canal would be overshadowed by the railroads emanating from Chicago, but the canal and lake trade helped start the city on its growth as the great inland lake port of the Midwest.

Even though Wisconsin and Michigan were lightly settled in the 1830s, canal surveys were part of the advance of the frontier, but most were for canals that would never be built. If the construction of canals at the portages elsewhere in the Old Northwest seemed inevitable, one may question why no canal ever successfully surmounted the historic Fox-Wisconsin River portage in Wisconsin, which had fascinated Frederick Jackson Turner, the historian of the frontier. A private company was chartered to build such a canal in 1837, and work began the following year. But little was done. A congressional land grant in 1846 for this project accompanied Wisconsin statehood in 1848. Canal construction began again in 1849, this time as a state enterprise conducted under a board of public works until 1853. A small steamer was able to pass through a canal in 1856, but sustained operation proved impossible.

When land sales could not support the enterprise, it was given to a succession of private companies until efforts to develop a navigable through waterway were finally abandoned in 1886.[49]

Farther north in Michigan, however, the great portage at St. Mary's Falls became the site of a capstone canal, built in the last decade of the Canal Era. Less than a mile in length, the Sault Canal passed the rapids and falls of the St. Mary's River to open navigation between Lakes Michigan and Superior. This short canal might best be compared to the short canal around the falls of the Ohio at Louisville.

To build the Sault Canal, Michigan received a land grant from Congress in 1852 after years of frustrating delay, much like the delays that still prevented completion of a canal between the Fox and Wisconsin rivers. A new company headed by Erastus Corning of New York was awarded a contract to build the canal with support from land sales. The work was directed first by Charles G. Harvey, then more effectively by John W. Brooks. It was the company's good fortune to secure the engineering services of Captain Augustus Canfield of the Army Corps of Topographical Engineers, who had graduated from West Point in 1822 and had made a survey and plan for a canal at the Sault in 1839.[50]

Canfield died in 1854, before the canal was finished, and Brooks turned to John T. Clark and William J. McAlpine, who had given long years of service on the enlargement of the Erie Canal. Indeed, there were many interrelationships between the Sault Canal and the New York waterway. Both were segments of the long water route from Lake Superior to the Hudson. From the beginning, Canfield had been assisted by L. L. Nichols, an Erie Canal engineer. One of Corning's closest associates as a director of the company was John V. L. Pruyn, who had aided in the enlargement of the Erie Canal. Another was John Seymour, brother of Governor Horatio Seymour of New York, who was just then pushing enlargement of the Erie Canal forward decisively. Their father, Henry Seymour, had been one of the original acting commissioners on the Erie Canal and had been engaged in the construction of the Erie Canal the same year that Canfield had finished his engineering training at West Point.

The Sault Canal was finished in 1855, with two deep-water locks 350 feet long and 70 feet wide, at a cost of not quite $1 million. The state of Michigan had its canal, and in the bargain the St. Mary's Falls Ship Canal Company won rich mineral lands in the land grant, which provided the basis for speculation that deeply affected the development of northern Michigan. The major articles of commerce on the canal were copper and iron from the upper peninsula, the former reaching more than 9,000 tons and the latter swelling to 120,000 tons in 1860. But the Sault Canal, like those on the Ohio and Illinois rivers, would eventually become obsolete and would then be rebuilt.[51]

The canals of the Old Northwest were in some ways similar to those of the mid-Atlantic and South, but because of their great extent and their frontier environment, the western canals also reveal common themes and unique developments. Most of the canals of the Old Northwest were built by the states or by state-assisted companies. Yet like De Witt Clinton in New York, Mathew Carey and William Lehman in Pennsylvania, or Charles F. Mercer and Joseph C. Cabell in Virginia, there were redoubtable figures in the Old Northwest who made the cause of state canals their own. Alfred Kelley was the De Witt Clinton of Ohio. He came from New York to Cleveland and later moved to Columbus. As a canal commissioner, he chose the Cuyahoga Valley route for the Ohio and Erie Canal, displaying great integrity and political skill, and he personally supervised construction from Cleveland to Akron. As a Canal Fund commissioner after 1841, he developed his relationships with the Commercial Bank of Lake Erie, the Franklin Bank of Columbus, and the Ohio Life Insurance and Trust Company—all of which provided indispensable deposit and disbursement service for the Canal Fund Board. Kelley sold Ohio bonds in England, and in the critical year of 1842 he used his personal property to guarantee the interest on canal bonds. When he turned to railroads in the 1850s, some of them followed routes he had surveyed for canals.

The Ohio canals were almost a joint venture between Kelley and Micajah T. Williams of Cincinnati. Williams wrote the Ohio canal bill of 1823 and won support in Cincinnati for the initial plan for two canals. He surveyed the southern part of the Ohio and Erie Canal with equal concern for engineering and politics, and he surveyed the Miami Canal. His canal experiences led him into banking, and he organized the Clinton Bank at Columbus and was president of the Ohio Life Insurance and Trust Company, which bought canal bonds and helped finance both the Miami Extension and the Wabash and Erie Canal. His son Jesse L. Williams was a rodman on the Ohio and Erie Canal at the age of seventeen in 1825 and seven years later became chief engineer on Indiana's Wabash and Erie Canal. Michjah T. Williams expanded his interest to encompass the Old Northwest. He was in the Louisville and Portland Canal Company, which built the canal around the falls of the Ohio River, and he was president of the Cincinnati and Whitewater Canal, which linked the Whitewater Canal in Indiana with Cincinnati. He speculated in lands in Toledo at the terminus of the Miami Extension Canal, in the copper country being opened up by the Sault Canal in northern Michigan, and in Milwaukee, Wisconsin.

A comparable figure in Indiana was Samuel Hanna. A pioneer fur trader who had arrived in Fort Wayne in 1819, he led in the movement for the Wabash and Erie Canal. Like Kelley and Williams, he was both a canal commissioner doing his own surveys and a Canal Fund commissioner; he was associated with eastern banks and helped organize the State Bank of Indiana. A servant of public canals, he was also a merchant, speculator, and banker, and he moved with the times from canals to railroads.

Similarly, the fortunes of the Sault Canal in Michigan rested in the hands of John W. Brooks. He had come from Massachusetts, where he had studied engineering with Loammi Baldwin, Jr. As vice-president of the St. Mary's Ship Canal Company, he was a key figure in transferrng the completed canal from his company to Michigan in return for the mineral and pine lands of the national land grant that were reserved for his company. Like the others, his career embraced railroads as well as canals.

The canals of the Old Northwest were also linked in the training of their engineers, many of whom came from the Erie Canal in New York. The ubiquitous Benjamin Wright, the original chief engineer of the Erie Canal, made the final decisions on the plan for the Illinois and Michigan Canal in 1837. James Geddes found the critical line across the Licking summit in Ohio, which made it possible for the Ohio and Erie Canal to combine a northeastern and central route and terminate at Portsmouth on the Ohio River. David Bates of New York helped locate the Louisville and Portland Canal on the Ohio River, and Bates and Nathan Roberts of New York surveyed the line of the Miami Canal. They taught Jesse L. Williams, who taught the Ohioan Samuel Forrer, who worked on the Ohio and Erie Canal, both segments of the Miami and Erie Canal, and the Wabash and Erie Canal. On the Sault Canal, John W. Brooks turned to Erie Canal engineers after the death of the army engineer Captain Augustus Canfield.

The itinerant nature of canal engineering on the midwestern canals is evident in the family of Seneca Lapham. Lapham and his sons, Darius and Increase, came from New York to Ohio. Seneca, skilled in building stonework and lock gates, worked under David S. Bates on the Erie Canal, then moved to the Schuylkill Navigation in Pennsylvania, and then returned to the Erie Canal, where he helped build the Rochester aqueduct and the locks at Lockport. His two sons worked on the Erie Canal and came with their father to Ohio to work on the Miami Canal, where Increase worked under Byron Kilbourne and Samuel Forrer. But Seneca and Increase went to work with Bates on the Louisville and Portland Canal before they returned with Bates to the Ohio and Erie Canal. Seneca remained on the Ohio and Erie Canal until 1830, while Increase lived at Columbus and worked on the canal under Francis Cleveland near Chillicothe until it was completed. Darius became assistant engineer on the Ohio and Erie Canal in 1829 and then superintendent on the Miami Canal. While in the latter position he designed the Cincinnati and Whitewater Canal. He became collector of tolls in Cincinnati in 1848 and died of cholera in 1850.[52]

Meanwhile, the Laphams met other canal engineers. William Gooding met Darius and Increase at Carroll, Ohio, before Gooding moved to the Wabash and Erie Canal and became chief engineer on the Illinois and Michigan Canal. While the Laphams worked on the Louisville and Portland Canal they were visited by Canvass White and George T. Olmstead from the Erie Canal, William Jerome from the Champlain Canal in New York, James Bucklin from the Miami Canal (and later the Illinois and Michigan Canal), and Alonzo Baldwin from the Delaware and Hudson Canal. The activities of this family touched half a dozen canals, and they met other engineers who worked in New York, Pennsylvania, Ohio, Indiana, Illinois, and Wisconsin.

The neat lines that show canal routes on maps seldom reveal the problems their engineers confronted and the conditions under which their builders worked. Geddes was debilitated by his work on the Ohio canals, and disease took the life of Seymour Skiff, a New York engineer who came to Ohio to work in 1823. After fourteen days on the job, he died of malaria and was replaced by another New Yorker, William Price. Cholera took its heaviest toll in the early 1850s along the last hundred miles of the Wabash and Erie Canal, but in 1854 one-tenth of the seventeen hundred men laboring on the Sault Canal succumbed to the disease. Some were among the four hundred men who had worked at the Sault through the bitter cold of the previous winter, eleven and a half hours a day. John Dickinson, the historian of the Sault Canal, has described the conditions facing men who reported for work early in the morning and had to locate their tools, which might be buried under a blanket of snow, and then dig through two feet of ice to reach the rock which they excavated from the frozen ground.[53]

At the Sault, the shortage of labor in 1854 led to repeated recruiting trips to New York, where some seventy men were enlisted, though John Seymour, who directed the search, was informed that "the men hired in N.Y. are generally to [sic] weak and small for such heavy work." But he was also warned that "as the cry is for men! men! you had better continue to forward more," and he continued to recruit.[54] Studies of the Indiana, Ohio, and Michigan canals all show reliance on Irish and German labor, with some immigrants moving from canal to canal. Maurice Cody, an Irish immigrant, came to America in 1825, worked on canals in New York, Pennsylvania, and Maryland, and then moved to Fort Wayne in 1834 to dig canals in Indiana.[55] The 1850 census for Pike County, Indiana, listed Timothy Donovan as head of a household containing ninety-four canal laborers, all but three born in Ireland. The migration of canal laborers is indicated in census figures for some counties for households containing a canal laborer with wife and family. In these households most wives were born in Indiana but children over the age of fifteen were born in New York or Ohio.[56]

Conflicts between rival groups of Irish laborers occurred on the Wabash and Erie Canal between Fort Wayne and Huntington in 1835, perhaps continuing the conflicts that had begun on the Chesapeake and Ohio Canal. About six hundred armed Corkonians and Fardowners were massed against each other. They raided each other's camps and brought canal work to a halt. The militia was called out, sheriffs arrested the leaders, and open battle was averted. David Burr reported to Governor Noah Noble that many had come from the Chesapeake and Ohio Canal and had "engaged in those bloody affrays at Williamsport and the high rocks on the Potomac" in Maryland.[57]

Many of the contractors were also Irish immigrants. Catherine T. Tobin has identified several Irish contractors on the Illinois and Michigan Canal, men such as John Melody, who left Ireland in the early 1830s, moved briefly to New York, and then took a canal contract in Illinois. William Snowbrook came to New York from Ireland in 1826, became a contractor on the Morris Canal in New Jersey, took a contract on the Maumee side-cut canal near Toledo, and then joined with William B. Ogden to take contracts on the Illinois and Michigan Canal. A similar path was followed by Patrick Casey, who arrived in New York from Ireland in 1837, worked on canals in New Jersey until 1840, and then became a canal contractor in Illinois. The latter two contractors later became prominent businessmen in Chicago.[58]

Though less numerous than the Irish on the canals of the Old Northwest, German immigrants can also be identified among the canal laborers. Frederick Christian Brase was born in Germany in 1821; he arrived in Indiana in 1844 and worked on the Wabash and Erie Canal until 1850. Sometime before 1860 he became a canal boat captain, hauling grain to Toledo and Cincinnati.[59] And while Irish and German immigrants joined the labor force, native New Englanders were adding their leaven to the Western Reserve, to Columbus, to Cincinnati, and along the Wabash and Erie Canal in Indiana. Asa Fairfield came from Kennebunk, Maine, and built the first canal boat on the Wabash and Erie Canal.

The relationship of Nathan Rowley and his son-in-law Thomas D. Smyth illustrates the mixed origin of canal labor in the Old Northwest. Rowley was born in Vermont, settled in Evansville, Indiana, and became a canal contractor on the southern division of the Central Canal before that section was designated as part of the Wabash and Erie Canal. He lobbied for the Butler bill in 1846, built canal boats, and became collector of tolls at Evansville in 1852. Smyth came from Ireland in 1838, worked with Rowley on the canal, and operated canal boats between Evansville and Toledo from 1855 to 1858. In 1855 Smyth joined the Evansville Guards, who put down the Clay County vigilantes, and in 1858 he became a repair superintendent on the Wabash and Erie Canal. Both men had responsible canal careers and after their death were buried in Evansville's Oak Hill Cemetery near the banks of the canal on which they had labored.[60]

Much of the canal labor in the Old Northwest was found locally. Contracts were usually let to local contractors, and both skilled and unskilled labor was required. Just as the New York canal commissioners reported in 1823 that most workers on the Erie Canal "were born among us," a study of labor on the Ohio and Erie Canal concluded that "the principal source of canal labor, as may be expected, was found in the local populace."[61]

Characteristic of most canals in the Canal Era, but especially in the Old Northwest, canals were central to the lives of the people who awaited their completion and used them when done. A resident of a town in

Indiana on the Wabash and Erie Canal wrote in 1845, "Having lived for the last seven years in a community whose only sentiment and only hope was the completion of this canal every possible effort was made to hurry forward the work."[62] In the Old Northwest urban growth followed the canal routes. Established cities such as Cincinnati, Chillicothe, Cleveland, and Toledo in Ohio competed vigorously for canals and grew with the canals they won. New midwestern cities were created by the canals: Akron at the Portage summit, Columbus on a feeder, Dayton as a canal terminus, Milan as an inland lake port, Chicago as a city stimulated by a canal before it became a rail center, and in Indiana a long line of interior towns and cities. More than any other canal settlement in Indiana, Fort Wayne was transformed from an Indian trading center at a portage place into a canal town.[63]

Almost as soon as new canals opened, the canal villages, towns, and cities were linked by packet boat travel. In the Old Northwest passenger travel was made easier by the long distances covered by the trunk lines that were connected to the numerous branch canals and navigable feeders. These canals were interstate waterways, crossing the Ohio and Indiana border and passing from Ohio into Pennsylvania. Whether or not canal travel appealed to travelers, passengers were driven onto the packets by the poor roads on which mud, dust, and the dangers of stage travel were characterisic.

Passenger canal travel in Ohio peaked at 52,922 in 1842. The largest passenger traffic was on the Ohio and Erie canal, where some 20,000 people arrived at Portsmouth by packet in 1843.[64] Most passengers traveled on the canal packets, but others took passage on the line boats that carried both freight and passengers. Terry K. Woods has descrbed the emergence of "express packets" on the Ohio canals after 1837, when the Ohio Packet Boat Company was organized in Cleveland.[65] These craft ran on fixed schedules, limited the number of passengers to thirty or forty per boat, and often carried mail. They set their schedules to connect with lake and river steamers or boats at the junctions with other canals. Packets ran the length of the Ohio and Erie Canal from Cleveland to Portsmouth in eighty hours. Horses were changed every ten to fourteen miles, and way stations were maintained to supply fresh teams. But some packets had an indefinite departure time, leaving whenever the captain gathered a sufficient load. The fare on the early packets was five cents a mile, or about $15 for the journey from Lake Erie to the Ohio River.

The size of both packets and line boats was determined by the size of the locks, which had dimensions of ninety by fifteen feet and could accommodate boats of seventy-eight to eighty feet long and fourteen feet wide. Passing the locks delayed a canal journey and was made more tedious because Ohio did not employ regular lock tenders before 1837. Cyrus Bradley, a seventeen-year-old Dartmouth College student, traveled on an Ohio and Erie Canal packet in 1831. He remembered his travels on the long levels without locks on the Erie Canal in New York and complained of the "plaguey locks" in Ohio.[66] The profiles of the Ohio canals show how frequently the locks interrupted a smooth passage whenever the canal was raised or lowered. In eastern Ohio there were the numerous locks up to the Akron and Licking summits; in western Ohio there were locks up to the Loramie summit whether moving north from the Ohio Valley or south from Lake Erie. Every delay witnessed the same time-consuming procedure, to enter the lock, fill the chamber with water or release it, and move out onto the new level of the canal. Though packets took precedence over freighters, boats lined up, entered the locks alternately, and were subject to delays from accidents and damages to the locks. Damaged boats sometimes sank in the lock.

Packet travel was especially welcomed in frontier Indiana, and these boats opened communication in the sparsely settled northern part of the state. If the new packets were colorful and adventurous, they traveled through a prairie landscape often flat and unimposing. A young lady from Louisville, traveling in 1851, referred to the occasional villages as "forlorn looking places," but her greatest complaint was of the mosquitoes, from which there was no escape. In her "hot and stuffy little room" she wrote, it seemed that "all the mosquitos ever hatched in the mud puddles of Indiana were condensed into one humming ravenous swarm right around my hard little bed. . . . All night I lay there under a smothering mosquito bar and listened to the buzzing of the insects, perspiring as I never supposed that anybody could."[67]

An English country gentleman, J. Richard Beste, boarded an eastbound packet boat at Terre Haute in August of 1851 accompanied by his family of eleven children between the ages of two and nineteen. They were bound for Toledo, beginning their return to England. Beste's account described their craft, which was pulled by three horses, with its men's and ladies' "saloons" and its long upper deck piled with luggage, where "passengers walked up and down or sat to enjoy the view." They slept on shelves hooked to the walls, with mattresses and sheets laid upon them. On the advice of the Terre Haute physician who had attended him during an illness, Beste gave each of his children a tablespoon of brandy every morning and night "in the hope of keeping off the ague and fever of the canal" and he mixed whiskey with the canal water they drank. They suffered most from the mosquitoes that "bit us dreadfully." At Lafayette, Beste and his family moved to another boat and he had "sent on board one of those pieces of furniture which are found in every European bed-room, but not one of which exists in any boat on this canal." He was aware that English bondholders now held the canal through trustees and he wished that "the English shareholders may send out a supply" of such commodes. While he marvelled at the "scores and scores of miles of woodland that had never heard the axe" near Fort Wayne, he was annoyed by a passenger who stood all day on the upper deck "with a fowling piece in hand, and constantly fired at the birds that flew across the canal." Nearer Ohio the number of passengers increased until he "never saw people packed so close" as they slept in the men's saloon. After a canal journey of almost five days Beste and his family reached Toledo and departed "our hateful boat—for the wretched fare and accommodation on which I paid about forty-five dollars a head."[68]

Since the Wabash and Erie Canal terminated at Toledo and joined the Miami and Erie Canal at Defiance, much of its passenger traffic was linked to Ohio and other states. The first boat to reach Evansville in 1852 was the *Pennsylvania*, built above Pittsburgh, which came up the Miami and Erie Canal to Toledo and then west to Evansville. In 1852, the Toledo and Wabash Line received two new boats made in Utica, New York, one named the *Northern Indiana* and the other the *Southern Michigan*. Praising this line, an Indiana editor noted that "the travel has been immense, and must have paid handsomely. The boats have generally been well filled, and frequently to their utmost capacity."[69] Although on most canals freighters far outnumbered packets, for the decade after 1843, when the canal reached only as far west as Fort Wayne, the passenger mileage was three times as great as the freight mileage. In 1843 regularly scheduled packet lines ran 104 miles west from Toledo, and two years later they ran 242 miles from Toledo west to Lafayette, covering the distance in fifty-six hours. Packet fares here were as little as two cents a mile. But as the numerous waybills that have been preserved attest, there were many short trips from canal port to canal port, just as was true on other canals. On the Wabash and Erie and the Miami and Erie canals together there were some twenty thousand passengers a year as late as 1850. That year, packets on the shorter Illinois and Michigan Canal carried more than twenty-two thousand passengers from Chicago to La Salle and seventeen thousand from La Salle to Chicago.[70]

Charles R. Poinsatte has written that most immigrants who traveled inland in Indiana did so by all-water routes. They were most likely to travel on line boats on the canal, which were cheaper and slower than the packets. Three thousand passengers a year came to Fort Wayne by canal in the early 1850s, many of whom were English, Irish, or German immigrants. Moreover, a significant number of immigrants entered Indiana from the South, coming from New Orleans up the Mississippi and Ohio rivers and entering the Wabash and Erie Canal at Evansville.[71]

A picture of a canal boat was placed at the center of the Great Seal of the state of Ohio, and the bicentennial celebration of Cincinnati in 1988 featured the reproduction of a canal lock. The canals of the Old Northwest were especially regional waterways. The region's development took place when the West was at a newer stage of growth than the older societies along the Atlantic. Eastern canals had as their great purpose the surmounting of the Appalachian barrier, seeking the trade of anticipated western growth. Western canals borrowed eastern technology, used eastern engineers, and helped to develop the great trans-Appalachian region stretching to the Mississippi.

This was Frederick Jackson Turner's "New West," exhibiting much of his interpretation of the frontier process between 1815 and 1850.[72] When the Miami Canal opened in Cincinnati in 1827, a local editor revealed a self-consciously western identity associated with the canal when he wrote, "The social, but noiseless movement of half a hundred backwoodsmen . . . along the unruffled surface of an almost currentless river, across a region . . . but recently reclaimed from the possession of its aboriginal lords—would have excited wonder, or incredulity, in any age but the present."[73] The passing of the frontier is illustrated in the efforts of the Indian agent John Johnston to direct the Miami Extension Canal through his lands near Piqua when he served as Ohio canal commissioner from 1825 to 1836. Similarly, the home of William Henry Harrison, frontier hero of Tippecanoe, at North Bend in Ohio looked out on the Miami and Whitewater Canal. The passage of the Wabash Indians down the Miami and Erie Canal to Cincinnati, en route to a reservation farther west, signaled the loss of their lands in Indiana. And frontier lawlessness marked the vandalism that destroyed the Birch Creek Reservoir, where there was opposition to the extension of the Wabash and Erie Canal to Evansville.

In this New West, canals came hand in hand with the steamboats on the rivers and the Great Lakes. Before the opening of the Louisville and Portland Canal in December 1830, most steamboats on the Ohio operated below Louisville. Only the *General Pike,* with a cargo capacity of 176 tons, operated regularly between Louisville and Cincinnati until 1825, and it was replaced by the *Ben Franklin* in 1826. But in the first year after the Louisville and Portland Canal opened, four hundred steamboats and an equal number of other craft passed through the canal. The number of steamboats going through the canal rose to more than a thousand in 1835 and averaged thirteen hundred a year until 1851.[74]

Steamboat design evolved in the 1830s and 1840s to produce boats with long, narrow, flat-bottomed, straight-sided hulls, above which rose three decks for cargo and staterooms. After Henry Shreve's *Washington* moved boilers and machinery up the main deck in 1816, two engines powered double paddle wheels, and by the 1850s chimneys reached ninety feet above the water. Tonnage on these boats rose from the 176-ton capacity of the *General Pike* to the giant *Sultana* built in 1848 carrying 1,700 tons. Most steamboats of the 1830s and 1840s could accommodate 300 to 600 tons.[75] And the steamboat helped increase the downstream flatboat trade because crews could make an easy, rapid return upstream, and the number of flatboats on the Ohio peaked in 1846-47.[76] Steamboat development responded to the opening of the Louisville and Portland Canal and to the commerce of the canal-river ports of Evansville, Lawrenceburg, Cincinnati, Portsmouth, Marietta, and Pittsburgh. After 1855, steamboats on the Ohio River declined in numbers because of railroad competition.

At the northern termini of the Ohio-Indiana-Illinois canals, steamboats helped carry canal commerce to eastern markets on the Great Lakes. The *Walk-in-the-Water* in 1816 and the *Superior* in 1823 inaugurated steamboat service to Detroit from Buffalo and Black Rock on the Niagara River in New York. After the opening of the Erie Canal in 1825 and the development of harbor facilities at Buffalo, steamboats multiplied into the hundreds. Yet dramatic as the rise of steamboats may appear, before the Civil War most of the grain sent to the Great Lakes by canal continued its passage to eastern markets in the holds of sailing schooners rather than steamboats.[77] We are reminded of the increased number of flatboats downstream alongside the steamboats on the Ohio River into the 1840s; by contrast, though steamboats on the Ohio declined after 1855, those on the Great Lakes continued to multiply.

Whether carried by steamboat or sailing vessel, most of the grain going east had left the canal or lake ports of Chicago, Toledo, Milan, Cleveland, and Erie. John G. Clark has noted the ability to tap interior grain-growing regions as the most significant reason for the rise of these lake ports. Their ranking thus reflected the time of the completion of their respective canals and the interior grain production stimulated by these canals. Cleveland contributed most to the Great Lakes grain trade until 1854; Milan's shipments of wheat and flour reached half those from Cleveland in 1847 and then declined; Toledo overtook Cleveland in the volume of grain shipments by the late 1850s.

Not until 1855 did Chicago rival Cleveland and Toledo in the volume of grain shipped. Chicago's grain shipments reached more than 2 million bushels only a year after the completion of the Illinois and Michigan Canal in 1848 and the arrival of the first wheat by railroad in the same year. Only a third of the wheat came by canal in 1850, and the canal's contribution was more important in the rise of corn shipments, 90 percent of which came to Chicago by canal in 1851, when corn had become the most valuable commodity shipped from Chicago.[78]

At the eastern end of Lake Erie, the millions of bushels of grain arriving from the canal-lake ports constituted the major portion of the trade of the Erie Canal in New York, which had carried mostly local products in its first years of operation and had to be enlarged in 1835 to accommodate western grain. Grain shipments also passed through the Welland Canal in Canada, going to the Oswego Canal in New York or on to Montreal.[79]

The long canal routes between the Ohio River and the Great Lakes were attempts to reverse the pattern of trade from the Old Northwest to the New Orleans market and to send that trade by canal to New York City or Philadelphia. Some of this market shift was accomplished as lake ports reached far into the interior, to central Ohio, to the Muskingum and Wabash valleys. But for most of the Ohio Valley trade, New Orleans remained the primary market.[80]

Yet the midwestern canals contributed to a population shift toward the central and northern portions of their states, while at the same time the more densely settled southern sections supported the state canals that would serve to develop rival regions. John Barnhart described the Ohio

Valley as the "valley of demoracy."[81] Public support of state canals or state assistance to private canal companies was enhanced by this egalitarian spirit, notwithstanding the localism and state rivalry that caused contention over every canal. James Willard Hurst has described the acceptance of public planning in nineteenth-century American law, and state and national land grants contributed to public support for canals. Ralph D. Gray has suggested that Indiana chose the canal over the railroad for the Wabash route in this egalitarian spirit: any man could own or use a canal boat, but only a private company could operate a railroad. Egalitarian pressures brought demands for lateral canals and overbuilding, especially in the Loan Law of 1837 in Ohio and the Mammoth Bill of 1836 in Indiana.[82]

The Northwest Ordinance of 1787 had foreordained five large states in the Old Northwest and set the stage for long state-owned canals. These canals were western extensions of eastern waterways, and they were built by eastern engineers on eastern models. Yet they represented a canal-connected society given a linear extension greater than in any other region of the nation.

6

The Canal Network

The regional canals developed into an integrated, interconnected network, the full dimensions of which can be best grasped if seen as a whole. Though the completed canal system was an achievement of American modernization, the character of this network and the decisions to inaugurate or expand it rested on a traditional water-connected society. Settlements had long been made initially at the mouths of rivers, at the fall line, or at carrying places. The river was the route to the interior in the face of unbroken forest, inadequate roads, and difficult terrain.

Yet the rivers responded to the seasons and the weather, independent of human need or desire. Only rarely did they serve for transportation with the perfection idealized in God's gift of nature. Most often they were too high, too low, or too obstructed, with frustrating consequences for those who sought to use them. Limited, palliative improvement was demonstrated in the wing dams that gathered the water and deepened it enough to allow boats to shoot a stretch of rapids or escape the inevitable groundings resulting from a prolonged drought. Time-consuming toil at innumerable carrying places made it almost inevitable that canals would be considered as a way to extend the distance over which weight could be floated and recourse to wagons be avoided.

Americans could see in English, Dutch, French, and even Chinese examples what canals could do to improve rivers, and the eighteenth-century sense of order prepared the way for a rational adaptation of nature. The canal did not so much connect rivers as run parallel to them.

Canals improved on the rivers they paralleled, and they followed the great river valleys which determined places of settlement and offered tillable land for farming. The valleys used by the major canals were geographic watersheds, which gave identity to a region, induced a social cohesiveness, and were often places of great beauty. Thus the river valleys of the Connecticut, the Mohawk, the Susquehanna, the James, the Cuyahoga, and the Great Miami shaped the paths of the canals that flowed through them. In the upper reaches of such valleys, canals climbed to the limits of lockage and water supply, usually in a winding course, sometimes in a narrow gorge, often bordered by bluffs and cliffs.

Where land was more level, the canal could be straight as an arrow, with smooth, close-packed banks and a uniform depth. It could control a constant flow of water and take its water from the river it was designed to improve. Feeders carried water from river to canal, taking only the amount that was needed; waste weirs and guard locks carried the surplus safely off when the water ran high. Water-powered mill wheels foreshadowed the hydraulic construction that would turn a millrace into a machine for transportation rather than a source for power.[1] Perfected for the river, the long, narrow Durham boats offered a model for canal boats that could be raised or lowered in a lock or pass each other in a narrow canal.

Writing in 1934, Lewis Mumford coined the term *Eotechnic society* to describe one centered on water, built on wood, using horse and harness for motive power. Wood supplied the lining for a lock chamber bedded in soil or stone; wooden miter gates were operated by long wooden arms; wooden trunks carried the canal like a cradled river across an aqueduct; and wooden boats were fashioned by shipbuilders already experienced in building craft for rivers and seas. To minimize lockage, canals followed the contour of the land, turning on gentle curves, combining industry and nature in eighteenth-century harmony. As John Stilgoe has written, the canal engineers "made straight—made geometrical the paths of agriculture and artifice."[2]

The canal engineers were the first professional engineers in America. Daniel Calhoun has estimated that in 1816 there were no more than two engineers or quasi-engineers per state, about thirty men altogether.[3] They usually emerged from a background of surveying, but their canal engineering was based on British or Dutch experience. Darwin H. Stapleton and Elting Morison have examined the transfer of canal technology from England to America over the thin thread of personal relationships, reaching back to James Brindley, whose labor on the Duke of Bridgewater's canal set the standard for canals in America. Brindley's knowledge was shared by William Weston, who came to America to work on the Schuylkill and Susquehanna Canal in the 1790s, assisted George Washington on the Potomac navigation, and responded to Loammi Baldwin's plea for help on the Middlesex Canal in Massachusetts. Weston shared his leveling instrument along with his knowledge, and he joined Benjamin Wright in his struggle alongside Philip Schuyler to improve the navigation of the Mohawk. Weston taught Baldwin and Wright, and Wright taught James Geddes, who followed Brindley's maxim to avoid lockage as much as possible when he located a line for the Erie Canal in western New York. Weston also taught Robert Brooke, who became an engineer on the Pennsylvania canals.[4]

Benjamin Latrobe came from the European continent and England and joined Weston in pursuing the "Philadelphia plan" to improve transportation to the Susquehanna Valley and western Pennsylvania. He worked on canals on the lower Susquehanna and assisted Joshua Gilpin and his company in their search for a route for the Chesapeake and Delaware Canal across the Delmarva Peninsula. In the process, Latrobe trained two American engineers, William Strickland and Robert Mills. As Darwin H. Stapleton has described them, Weston and Latrobe established the precedent for professional direction of American canals and, even more, "provided the foundation for the rise of an American engineering tradition."[5]

On the pioneer canals in the South, John Christian Senf served as the link between the canals of Holland and American engineering knowledge. Senf was a Swedish military engineer who had served Washington during the American Revolution and had visited the canals of Holland. He assisted Washington on the Potomac navigation but was best known for his work as the major engineer for the Santee Canal in South Carolina.

The Erie Canal became the great training ground for canal engineering. Wright and Geddes laid out the greatest portion of the canal line. Canvass White made a trip to England to observe the use of Roman cement and then developed in New York the underwater cement that bound the stone structures for virtually every subsequent American canal. David Bates located the line of the canal along the Mohawk in its descent through that beautiful valley; Nathan Roberts designed the dramatic five-lock flight up the Niagara escarpment at Lockport; and John B. Jervis of Rome launched an engineering career that would lead to the building of the Croton Aqueduct in New York and the invention of the movable truck on American railroad engines.[6]

Even before the great New York waterway was completed, New York engineers exported their skills to other canals. They were itinerant bearers of a new technology influencing the nation much as New England influenced American cultural life during the same period. As chief engineer on the Erie Canal, Benjamin Wright attained a reputation greater than that of any civil engineer in America. In 1824 he became chief engineer on the Chesapeake and Delaware Canal, where he became involved in a bitter and costly dispute with another New Yorker, John Randel, Jr. A decade later, Wright became chief engineer of the James River and Kanawha Canal in Virginia and would be similarly embroiled in a conflict with Charles Ellet, Jr., but his reputation still enabled him to have a decisive impact on other projects such as the Illinois and Michigan Canal. Other New York engineers turned to him as a consultant, and he worked with Canvass White on the Union Canal in Pennsylvania, with John Mills on the Delaware and Hudson Canal, and with Ephraim Beach on the Morris Canal in New Jersey.[7]

Almost every canal built during this era experienced the influence of a New York engineer. Canvass White became chief engineer on the Union Canal in Pennsylvania and then worked on the Lehigh Canal in Pennsylvania and the Delaware and Raritan Canal in New Jersey. James Geddes located the line on the critical portion of the Ohio and Erie Canal; he made the first design for the Allegheny Portage Railroad that surmounted the mountain divide on the Pennsylvania Mainline Canal, located the Lehigh Canal to Mauch Chunk, and surveyed the Delaware Extension Canal from Easton to the Chesapeake. Nathan Roberts located the Miami and Erie Canal, which passed through the Black Swamp in northwestern Ohio, not unlike the Montezuma Swamp in New York, and with Geddes he helped train the Ohio engineer Samuel Forrer. Jarvis Hurd helped design the Massachusetts part of the New Haven and Northampton Canal. Wright, Roberts, and John Martineau were on the Board of Engineers of the Chesapeake and Ohio Canal. The success of these engineers bred an institutional, organizational loyalty, apart from their individual talents, creating a new type of American engineer. In Troy, New York, civil engineering became part of the curriculum of the Rensselaer School in 1831. Raymond H. Merritt has traced the careers of the graduates of the Rensselaer School who were the first to receive civil engineering degrees in the United States.[8]

These engineers superseded the more individualistic style of the Baldwins, father and son, and James Sullivan in Massachusetts or Benjamin Latrobe and Robert Mills in the South. Thus Loammi Baldwin, Jr., carried his skills from the Middlesex Canal to other waterways much as did Benjamin Wright. He preceded Canvass White on the Union

Canal and Wright on the James River and Kanawha Canal. In each case, his career ran afoul because of his intense individualism, and he lacked the loyalty to larger organizations exhibited by the New York engineers.

The immense demand for engineers in the Canal Era far exceeded the supply trained on the Erie Canal in New York or under the engineering corps of the United States Army. Pennsylvania built more miles of canals than New York in more difficult terrain, at far greater expense. Pennsylvania engineers, who learned as they worked, soon outnumbered the New Yorkers on Pennsylvania canals. The Philadelphian William Strickland was trained by Benjamin Latrobe, and his meticulous report in 1824 for the Pennsylvania Society for the Promotion of Internal Improvements led to his initial surveys for the Pennsylvania Mainline Canal. The Pennsylvanians Solomon Roberts, James D. Harris, and William B. Foster overcame challenges of mountain canal building unlike anything faced in New York.

Josiah White was more entrepreneur than engineer, but his innovations in dam building on the Lehigh River and his efforts to canalize that tumultuous waterway anticipated twentieth-century large canal construction. The young engineer Edwin A. Douglas carried out Josiah White's plans for canals on the upper Lehigh and designed the highest lift lock at that time in America, with a lift of thirty feet. Ashbel Welch followed in the steps of his brother Sylvester, who came from the Erie Canal to the Lehigh Canal. In his own right, Ashbel Welch designed the long Delaware and Raritan Canal feeder, which ran parallel to the Delaware River, and he later rebuilt the locks on the main canal, which had been designed by Canvass White.

Long after the Erie Canal engineers had left Ohio, Samuel Forrer, Richard Howe and his grandson George A. Howe, Francis Cleveland, and the Lapham family of engineers continued work on Ohio's long canals. The young Jesse B. Williams began with Erie Canal engineers in Ohio but moved west to serve as chief engineer on the Wabash and Erie Canal in Indiana. Beginning his work in Indiana, James Gooding became chief engineer on the Illinois and Michigan Canal carrying the science of canal building to a river flowing into the Mississippi.

The superintending engineers directed the work of contractors who took a mile, a little more, or a little less, and worked to meet specifications established for their contract. Payment depended on the character of the earth or stone to be removed or the skilled construction to be done.

The contractors on the first canals were frequently farmers or artisans who carried their general skills into their new work.

In time, contractors became specialized entrepreneurs. Harry N. Scheiber has described the "professional canal contractor" who migrated from state to state, much as did the canal engineers.[9] Many contractors came to Ohio from the Erie Canal in New York and also from the Chesapeake and Delaware Canal. On the Blackstone Canal in New England, Benjamin Wright brought canal contractors with him from the Erie Canal, men whom Vincent Powers has described as "veteran" Irish canal builders. Powers demonstrated that many of the Irish immigrants on the canals were "expectant entrepreneurs" rising in their positions on the canal and improving their position in society.[10] Catherine T. Tobin identified successful Irish contractors on the Illinois and Michigan Canal and traced their rise in Chicago business and politics.[11] On that canal there were also professional contractors who moved from state to state. Hart L. Stevens, for example, testified in 1847 during an inquiry that he had been a contractor since 1824 on the canals or public works in New York, Ohio, Pennsylvania, and Indiana before coming to Illinois. Alongside these professional contractors, however, there remained the individual farmer or small businessman who engaged in work on the canals with little previous experience.

Peter Way has called the contractors "the linchpin in canal construction" on the Chesapeake and Ohio Canal. They needed a small working capital and were expected to have references attesting to their experience. They were paid on the basis of competitive bidding, the estimates of the engineers, or the allocations for the work to be performed. But in the competition for contracts there was a tendency to underbid. For example, there were almost five hundred proposals for the first thirty-four sections let on the Chesapeake and Ohio Canal.[12] Contractors had to procure laborers, feed and house them, and bear the risks of weather and floods. They entered frequent claims for damages caused by conditions that made it difficult for them to complete their jobs on the terms they had agreed upon.

Monthly payments were usually made based on the estimates of the engineers, though this accommodation could be suspended in times of economic stringency. Abandonment of contracts was most severe when the canal commissioners or companies lacked funds to disburse, especially for the canal projects under way during the late 1830s and early

1840s. In the close connection between the state banks and the canals, the revenue from loans, tolls, or other sources was deposited in the banks, which then paid contractors in their own bills.[13]

Few canal projects were without accounts of contractors who were unable to continue their contracts. On the ill-starred New Haven and Northampton Canal, Robert Sheldon worked until his considerable personal fortune was exhausted on his part of the line. On the Chesapeake and Ohio Canal, Lee Montgomery labored against such odds, in the midst of violence and destruction by the Irish laborers on the Paw Paw tunnel, that he was bankrupted in the process. Elias Cozad fell into debt on the Ohio and Erie Canal in 1827 and appealed to Acting Commissioner Alfred Kelley for another contract after having "spent two years and five months of my time labouring faithfully and unceasingly." He wanted one of the sections abandoned by other contractors. Harry N. Scheiber estimated that 12 to 15 percent of the contracts on the Ohio canals were abandoned by the contractors.[14] To forestall abandonment of contracts on the Wabash and Erie Canal as the state neared default in 1839 and 1840, Indiana paid contractors in "White Dog" or "Blue Dog" scrip. In Illinois, scrip paid to contractors during the Panic of 1837 declined to a third of its value by 1842. In Pennsylvania, Isaac Harris, a contractor in Pittsburgh, found himself "idle, poor [and] penniless" in 1832 after two years of canal construction; because "the state became embarrassed," he wrote, he had gone into debt and believed that the state owed him some $20,000. Many contractors, like Harris, blamed their losses on canal commissioners, superintendents, or politicians.[15]

The political scandal that involved embezzlement by Myron Holley and his payments to contractors on the Erie Canal in New York revealed the difficulties in disbursing large sums under difficult conditions. Holley served as a dedicated canal commissioner, moving up and down the Erie Canal paying contractors in small notes drawn on local banks and establishing close personal relations. He appears to have been honest and used careful accounting in his disbursements, but his career collapsed in scandal when he used state funds for personal land speculation. But Holley pleaded unsuccessfully for a remuneration of 1 percent of the $2.5 million he had disbursed as "an adequate and reasonable compensation for his services and hazards as treasurer" to clear his accounts of shortage.[16]

In 1846 a fat volume on canal frauds was published in New York detailing malfeasance in construction as well as political favoritism in letting of canal contracts.[17] In the period of the stop and tax law suspension after 1842, damage claims from contractors on suspended contracts were so great that the Whigs parodied the Democrats for paying "MILLIONS FOR DAMAGES BUT NOT A CENT FOR IMPROVEMENT."[18] When New York enacted the "Nine Million" bill to speed the enlargement of the Erie Canal in 1851, a legion of contractors descended on Albany for political advantage, and some of their work on their contracts appeared profligate.

Canal construction was made more difficult because many canals were built and then rebuilt to larger dimensions while the original canal continued in partial use. Much of the rebuilding took place in the depression years after 1837, when canal work was delayed by lack of funds, construction was sporadic, and navigation suffered. New construction proceeded amid the congested traffic on the old. The finely cut and fitted stones set in the locks and aqueducts that still stand today in state after state demonstrate a skill in construction well beyond what might be expected from the piecemeal, decentralized, often underfunded technology of the time.[19]

Conditions of canal labor were arduous in the extreme, though other laborers in the period endured equal hardships. Canals were built by a preindustrial labor force, not yet forced into the discipline of the factory or the mechanical regimens required by the machine. Although the direction of work on a canal was in the hands of canal commissioners, a board of public works, and the engineers, it was the contractors and subcontractors who managed most of the laborers.

Hours were from sunup to sundown, with stops for meals and "drams," or "jiggers," as a supply of alcoholic beverages was often stipulated in the labor agreements. Some states, notably Ohio, Illinois, and, for a time, New York, prohibited the use of alcohol on the job, but with only limited success. The enormous collection of pay vouchers in the New York State Library, saved from accidental loss to a Ticonderoga paper mill by a fortunate stroke of historical preservation, show average wages for labor on the Erie Canal and its enlargement at $12 a month or fifty cents a day. Monthly wages varied slightly from canal to canal, $12 or $13 on the Morris Canal, $14 on the Pennsylvania canals, $10 to $13 on the Indiana canals in the early 1830s and $18 to $20 in the 1840s, and an average of $20 on the Illinois and Michigan Canal for the entire period of its construction. Wages varied for skilled and unskilled labor. Unskilled canal labor in Ohio received $8 to $10 monthly in 1827 while in 1828 a carpenter was paid $21.[20]

Competition rapidly drove wages upward. Unskilled workers in Ohio were paid $13 a month in 1828, $16 in 1829, and as high as $18 in 1832. Wages for unskilled labor at the opening of construction on the Chesapeake and Ohio Canal in 1828 were $10 to $12 a month, but the next year common laborers were being paid $1 to $1.25 a day. Workers received board and housing along with their pay but were not paid for days of sickness, work stoppage, or bad weather. Geoge Rogers Taylor found these wages comparable to those in agriculture or textile mills during the Canal Era. Yet Peter Way has written that wages on the Chesapeake and Ohio Canal, with deductions for sickness, weather, or stoppages, generally covered twenty-six days a month and sometimes only sixteen or seventeen. Under such conditions, the workers "crawled along the razor edge of subsistence."[21]

The early years of canal building in the first half of the nineteenth century were a time of labor shortage, and the boom in canal building by states and private companies required the work of thousands of men. In 1821 there were nine thousand men at work on the Erie Canal. There were more than three thousand at work on the Chesapeake and Ohio Canal in 1829, the number rising to as many as five thousand in the next decade. More than two thousand worked on the Illinois and Michigan Canal in 1838, and in Indiana public works projects employed more than five thousand men after 1836. This great need meant ready employment for unskilled immigrants, and competition for labor was intense, pushing wages up. Yet when Avard L. Bishop searched through the reports of the Pennsylvania canal commissioners, he found "that at no time was there any serious interruption of work" or delay "caused by the scarcity of laborers."[22]

Irish immigrants supplied much of this labor, but other ethnic groups such as the Germans and the Welsh worked in large numbers on the canals. The New York canal commissioners, reporting on the construction of the Erie Canal in 1819, noted that three-fourths of the laborers were "born among us." A study of the Ohio and Erie Canal concluded that the "principal source of labor was found in the local population."[23] Convict labor was used in both New York and Illinois, and slaves were used on the southern canals. After 1822 immigrants began to replace native laborers on the Erie Canal, and President Charles

F. Mercer wrote in 1830 that among the laborers on the Chesapeake and Ohio Canal, "the greater part of them are transient foreigners; sometimes on the Pennsylvania Canals; sometimes on the Baltimore & Ohio Railroad; sometimes at work on our canal."[24] It is difficult to ascertain what proportion of the canal laborers were Irish immigrants, but they appeared so regularly on American canals that they have been studied in detail.

Active recruitment of Irish labor abroad appears to have been very limited. Charles F. Mercer of the Chesapeake and Ohio Canal Company contacted the American consuls at Cork, Belfast, and Dublin and the American minister in London seeking laborers but with little result. Irish workers were sent to American canals by the Irish Emigrant Association and the Union Emigrant Society. Once in America, the kinship connections among the Irish brought others.[25]

The first Irish to come to American canals were those who landed in Canada and then moved to New York to work on the Erie Canal. Irish immigrants arriving at eastern ports found numerous advertisements for work on canals, especially in the *New York Truth Teller.* Between eight hundred thousand and a million Irish immigrants came to North America between 1815 and the great famine of 1845, and in the decade after that terrible event another 1.8 million came.[26] The Irish immigrants brought with them their own cultural identity and worked in pre-industrial patterns. Kerby Miller has written that for the period 1825 to 1832, about half the Irish immigrants to America were from Protestant Ulster, but after that they were mostly poorer Catholic tenants and laborers of the southern counties in Ireland.

Most of the Irish who worked on American canals were unskilled. By 1836 only 3 percent of the Irish arriving at New York were classified as artisans or professional men, and most of the rest were laborers or servants. About two-thirds were young men, single, and in their early twenties. Many came out of the desperation of poverty and some for "independence" in the New World. Those who came after the famine were still poorer, even less skilled, and more dependent on immediate employment after arrival. Among those who worked on American canals, the immigrants who came after the famine were more likely than the earlier group to be guilty of drunkenness, violence, or crime. The first official literacy statistics in 1841 showed rates of 40 to 44 percent illiterate in Ulster and as high as 85 percent illiterate in Galway, Mayo, and Donegal.[27]

Their cultural traditions shaped Irish identities as Ulstermen, Corkonians, or Fardowners, the latter called Longfords. Corkonians were from southern Ireland and Fardowners were from central Ireland, and much of their conflict as they labored together on American canals was attributable to their rival county origins. David Grimsted has described the Irish labor riots on the canals as "imported clan battles between groups of Irish Catholics from different areas of the old country."[28]

Peter Way has emphasized the Irish secret society as the precedent for the response of Irish workers to the conditions of labor on American canals. The secret society, he wrote, "acted as their model for organization." Clan divisions turned Irish laborers against each other, often with threats, beatings, killings, and destruction, but ethnic bonds "mediated Irish workers' experience." At bottom, the affrays among Irish laborers were "mad scrambles for job security nurtured by an oppressive climate of impending unemployment." Laborers living on the edge of subsistence used coercion to keep other Irish (or Germans or native Americans) from taking their livelihood. On the Chesapeake and Ohio Canal, a shared ethnic identity was organized through the secret society and used threats and brutality to gain effective control over works such as the Paw Paw tunnel to force contractors to make wage payments and give regular employment. These conflicts contained elements of such modern concepts as the closed shop, syndicalism, and class conflict.[29]

Irish immigrants took up a labor on American canals that was physically demanding, monotonous, often tormented by heat, wet, and mud, and sometimes dangerous. Their basic tools were the shovel and the wheelbarrow, which they used in digging the canal channels, forming the banks, lining them with clay, and building towpaths. Technological advance and invention eased some of these tasks when tall trees were brought down by a machine using a roller, cable, and crank. A great wheel spoked to a pulley was developed to cradle a giant stump puller. Scrapers drawn by horses were perfected to reduce digging with shovels. Blasting of rocks was aided by the new company of E. I. du Pont de Nemours, which sold powder for the purpose.[30]

The workers pushing the line forward labored on canals often located in remote areas, which were thinly settled and far from the amenities of towns. They lived in shanties, shared by fifteen to twenty men under one roof with little protection from the elements. Families lived in rude, temporary huts. Irish laborers often worked in water, stood knee deep in mud, and dug canals through swampy lands while tormented by mosquitoes. Canals followed close by the rivers, and workers blasted a path through rocky cliffs and narrow gorges. Danger waited under heavy timbers and falling stones, and the rock blasting threatened unprotected eyes and ears.

Sickness was endemic. Cholera struck the Irish laborers along with others, especially in 1832, when the Chesapeake and Ohio Canal was hard hit. The epidemic of 1852 was especially severe on the Wabash and Erie Canal. Virtually every year canal laborers were struck down by the "ague" or "autumnal fever," and there were frequent outbreaks of malaria. The malaria epidemic of 1838 on the Illinois and Michigan Canal took between seven hundred and a thousand men.[31]

The Catholic priests rendered what help they could, especially Father John Blaise Raho on the Illinois and Michigan Canal. He wrote to Bishop Joseph Rosati, "The diseases in this area are horrible and so many die that there is hardly any time to give Extreme Unction to everybody. We run night and day to assist the sick." Father Raho responded to cholera much as the Protestants did, seeing the epidemics as the Lord's punishment of the workers for "getting drunk all the time, their riots, their fights and homicides."[32] Yet the Irish gave generously to their church when working on the Illinois and Michigan Canal, the Miami Canal, and others, and their presence contributed to the expansion of Catholic parishes.

The Protestant American Home Mission Society gave similar assistance on the Illinois and Michigan Canal. In 1845 the Reverend Hutchins Taylor reported from his station at Joliet on "the afflictions of Divine Providence upon myself and this people among whom I labor in the form of distressing sickness." He also reported on the loss of "much valuable time" to the laborers and the cost of "medicine at the Apothecary and the doctor's Bills," which were "almost overwhelming."[33] Without the modern protection of sick time, the canal laborer was usually paid only for hours actually worked.

Alcohol was part of an Irish laborer's life, whether from his native traditions, an antidote to disease, or a response to work on the canals. William J. Rorabaugh has described Irish drinking in the context of their hard work, menial jobs, and life "in an alien and sometimes hostile land"; they "reduced their tensions with strong drink, which they consumed exuberantly and in great quantity."[34] Their drinking was closely associated with their fighting, and it contributed to their reputation as the "fighting Irish."

Drunkenness and fighting among Irish laborers appeared spasmodically on American canals. The armed conflict among Irish workers on the Erie Canal at Lockport in 1824 involved Protestant Ulstermen and Irish Catholics. Excessive drinking among the Irish on the Chesapeake and Ohio Canal led to the company's enforcement in 1832 of a provision denying liquor to the laborers on the job, which only led to intoxication and fighting at night. This was followed by armed conflict between the rival Catholics, the Corkonians and the Fardowners, in 1834 near Williamsport, part of hostilities that continued through the 1830s and included the strike of 1837-38. Conflict between the Corkonians and Fardowners on the Wabash and Erie Canal in 1834 and 1835 near Huntington nearly became a pitched battle. The worst riot on the Illinois and Michigan Canal between the Corkonians and the Fardowners broke out in 1838 near Ottawa and La Salle and ended in a battle between the Irish and a local posse and the death of ten Corkonians. A labor riot took place on the Savannah, Ogeechee, and Altamaha Canal in Georgia in 1829, and a strike stopped work on the Pennsylvania canals in the same year. In 1834 an Irish labor riot occurred on the Chenango Canal in New York. Some of these conflicts involved Irish secret societies, as did one on the Delaware and Hudson Canal in 1827 and 1828 and those on the Chesapeake and Ohio Canal in the 1830s.[35]

Behind many of these conflicts, however, were real grievances over irregular payments, fraudulent practices by contractors and subcontractors, and demands for higher wages. The riots of 1834 and 1835 were part of a larger pattern of violent upheavals in the nation, which were identified by Carl E. Prince as part of a great "riot year." In his view, canal riots were "all of a piece with other episodes of violence, involving race riots, anti-abolitionist rioting, and election riots." All were challenges against "the wall of American community, as a pre-industrial society accelerated toward major change."[36]

The canal was a mechanical structure, deceptive in its appearance as a replicated river that was subject to the same natural forces as natural waterways. Yet the route of any canal was a man-made series of levels, which were determined by the contours of the land, with the new horizontal elevations or lower levels marked by lift locks that raised or lowered boats only a few feet at a time. But the rate of successive lockages, one every three or four minutes or as many as 250 boats a day on the Erie Canal, limited the efficiency with which a canal-machine could be operated, as did a series of other variables.[37]

Each lockage used a lockful of water, and the engineer-designers had always to calculate the available water supply, plan for subsidiary waterways to serve as feeders, and provide a minimal pitch for a sustained water current. They had to determine the most effective ratio between the contoured distance and the number of locks required to reach the desired destination, within the limits of available funds. Placid waters reflecting canal-side foliage belied the complex forces governing canals, which made their apparent order subject to uncontrollable changes in rainfall, drought, leakage, and flood. Breaks in the banks were the Achilles' heel of the canals. If the banks were inadequately lined or puddled, if muskrats dug tunnels or water seeped around culverts, a small section might give way, enlarge itself, and drain much of a level before it could be dammed.

Deceptive, too, were the frequent stretches of American canals requiring only mile on mile of excavation, such as the long levels of sixty and eighty miles on the Erie Canal in New York; the broad, flat stretches along the Delaware in Pennsylvania; and especially the minimal lockage on the 450 miles of the Wabash and Erie Canal in Indiana. John Stilgoe has described an American landscape ripe for canals, noting that the longest canal in Europe was not half the length of the Erie Canal.[38] Although the English model for canals was transferred to America by William Weston and Benjamin Latrobe and by the American engineers who visited England, it was undersized for the American experience. The small boats that plied English canals were inadequate for the Pennsylvania and New Jersey canals where such small boats were used most extensively. Darwin H. Stapleton has written that by 1816 "an American engineering style was emerging, one which . . . appears to have been better adapted to the American landscape and society than English engineering."[39]

Louis M. Hunter has described the new science of hydraulic engineering that was pioneered on the short canals at Lowell on the Merrimack River, in which the development of waterpower machinery paralleled the development of canal transportation.[40] But the Lowell application was only the most famous of the thousands of wheels turned by surplus waters on American canals. Any steep drop in canal levels offered waterpower to be exploited by the mills that sprouted along every canal and sometimes competed with the canal for the water.

Nonetheless, on almost every American canal the search for levels and an adequate water supply required engineering feats often described as stupendous. The vertical dimensions of American canals can be seen in canal profiles, which, for example, convey the elevation of the Ohio canals to the Akron summit, the Licking summit, and the Loramie summit, or the elevation of the Pennsylvania Mainline Canal in the valleys of the Juniata and Conemaugh rivers.

The achievements in canal construction make especially superficial any view of the Canal Era as an ephemeral interlude between the turnpike and the railroad. On the Erie Canal such feats were the mile-long Irondequoit embankment near Rochester and at Lockport the five-lock double combine rising sixty-six feet up the Niagara escarpment followed by the mile of deep-cutting blasted twenty-five feet wide through solid rock. On the Pennsylvania Mainline Canal such a feat was the Allegheny Portage Railroad at Hollidaysburg on which canal boats were carried in sections by stationary engines up almost fourteen hundred feet before being lowered to the level of Johnstown. On the Chesapeake and Ohio Canal, nearly impossible construction was finally thwarted in the mountains along the Potomac River near Harpers Ferry; and the James River and Kanawha Canal attempted the impossible above Buchanan on the Blue Ridge. In Indiana the Whitewater Canal climbed five hundred feet in fifty miles but could not sustain its stair-step footholds against the floods of the Whitewater Valley. The engineering ingenuity of the Morris Canal was displayed deep underground in the elaborate iron turbines that powered the inclined planes on this unorthodox waterway.

The five canal tunnels in the nation, especially the Paw Paw tunnel on the Chesapeake and Ohio and the tunnel near Cincinnati on the Cincinnati and Whitewater Canal, bordered on hubris. If many of the lateral or branch canals were economically unproductive, they sometimes went where they could scarcely have been expected to go, as in New York, where the Black River Canal scaled the foothills of the Adirondacks, or the Genesee Canal followed the deep gorge of the Genesee River. The canals of the upper Susquehanna went into operation only after prolonged construction which approached futility.

To carry the canal level and its water across rivers, the aqueducts were often like the Roman aqueducts of old, combining mass, length, height, strength, and beauty. Piers set in a rushing stream, stone cut for close fitting and bonded with cement, magnificent Romanesque arches supporting a towpath on one side and the wooden trough of the canal on the other, all made for an effect of majesty. Two such aqueducts crossed and recrossed the Mohawk on the Erie Canal. An aqueduct across the

Seneca River on the enlarged canal had thirty-one arches, and the aqueduct across the Genesee at Rochester was eight hundred feet long, resting on eleven arches. Those on the Chesapeake and Ohio Canal were among the most beautiful and best constructed in the nation, as was the aqueduct near Princeton on the Delaware and Raritan Canal. When John Roebling suspended his aqueducts by woven iron cables to cross the Allegheny River at Pittsburgh, the Delaware River on the Delaware and Hudson Canal, and the Potomac on the Chesapeake and Ohio Canal, he foreshadowed a new era in the use of suspension bridges in America. If the Ohio and Indiana canals were less likely to have aqueducts resting on Romanesque arches, some such as the Duck Creek aqueduct on the Whitewater Canal were roofed in the familiar design of the covered bridge. The Aboite aqueduct on the Wabash and Erie Canal used a broad platform of wood, which underlay the towpath and the trunk of the canal.

When such structures and levels were finished on a long canal line, the occasion was cause for celebration, especially in a canal town when water reached it, or for an entire waterway, often a celebration on the Fourth of July reflecting the self-conscious nationalism of the period. Wherever canal settlements grew there were ubiquitous basins, quays, and snubbing posts. Basins replaced corners in local place names, and inland towns added the suffix of "port" to their locations. In the canal basins the long, narrow canal boats could be turned around, loaded, and unloaded. Warehouses appeared, and in time grain elevators might be added. Rochester on the Erie Canal had twenty-one mills along the canal in 1835. There were so many mills on the Susquehanna Canal that they increased the current in the canal and caused it to silt up. When the canal basin served a town, hotels, taverns, and country stores appeared to service a transient public.

The passing of commerce and people linked canal towns and cities to an extended network. The long lines on a canal map appear as an interconnected system. The Erie Canal followed the Mohawk gap in the Appalachian chain across the state of New York to Lake Erie, and lateral canals reached north to Lake Ontario and south to link with the Pennsylvania lateral canals fingering north from the Pennsylvania Mainline Canal. The Pennsylvania canals reached south into the region of the Chesapeake and east to connect to the canals of New Jersey. The Chesapeake and Delaware connected the two great bays of the mid-Atlantic coast, and the Chesapeake and Ohio extended from Washington up toward the Ohio River as far as Harpers Ferry. No great trunk lines to the

West were completed in New England or the South, but canals through the Dismal Swamp offered a north-south inland water highway, and in the South fingers of canals reached from the uplands eastward to the sea.

This was a "long-distance communications network," as Richard D. Brown has observed in tracing the process of modernization in America, and it was also the beginning of the standardization of public works. Christopher T. Baer has noted that on the through trade on the mid-Atlantic canals, coal boats could go down the Lehigh Canal, down the Delaware Division Canal to the Chesapeake, over the Chesapeake and Delaware Canal to the Morris Canal in New Jersey, and arrive at the doorstep of New York City. Decked canal boats could be towed on Chesapeake Bay, the Hudson River, or Long Island Sound. Earl J. Heydinger has traced the canal traffic moving on a north-south line from the New York lateral canals to those of the upper Susquehanna branches, to the Susquehanna Division and the Mainline Canal, to the canals of the lower Susquehanna, and finally to the Chesapeake. Out of Chesapeake Bay to the south the Dismal Swamp Canal and the Albemarle and Chesapeake Canal reached south to Albemarle Sound.[41]

In the Midwest, the Pennsylvania canals linked with those of Ohio. Two trunk routes connected Lake Erie and the Ohio River, and lateral canals in Ohio reached into every section of the state. Indiana's great Wabash and Erie Canal joined the Miami and Erie Canal near Toledo and dropped in a long diagonal to the Ohio River, just above the Mississippi. A list of boats passing Fort Wayne showed home ports on the canal line in Indiana and on the Ohio section of the line to Toledo, but boats also came from Akron, Cincinnati, and even Buffalo, New York.[42] Boats from the Ohio canals were operated by companies that took them across the eastern end of Lake Erie to the Erie Canal and down the Hudson River to New York. The New York and Ohio Line in 1835 advertised boats "running day and night on the Erie & Ohio Canals."[43] Where the Ohio and Pennsylvania canals intersected at their state border, parallel routes ran to Cleveland and Erie, and Pennsylvania's canals offered a shorter route to the Ohio River at Beaver from the Ohio and Erie Canal than did the latter waterway where it connected to Marietta or Portsmouth.[44] Copper ore came down the lakes by steamer, entered the Ohio and Erie Canal at Cleveland, and was carried by canal to Pittsburgh.

In northern Indiana, boats left the Wabash and Erie Canal, joined the Miami and Erie Canal and turned northward toward Lake Erie at Toledo or south to Cincinnati on the Ohio River. And in southeastern Indiana, the Cincinnati and Whitewater Canal drew much of the trade of Indiana's Whitewater Canal to Cincinnati.

Yet for such an interlocked transportation system, the canals were built in the midst of conflicting interests. Many communities supported canals only if they would stop with them, and there was usually bitter rivalry over canal routes. As in the railroad era, some canals were built for no other purpose than logrolling and lacked a sustained means of support or survival. Rate discriminations were so blatant as to violate the constitutional prohibition of interstate tariffs.[45] Since most canals were state-owned, better integration of the system might have been expected. And if it were true that an entire canal must be completed for any part of it to have its fullest use, so too no canal could develop its fullest potential completely independent of the others.

Through travel by canal was sometimes impaired by variations in dimensions similar to the different gauges on early American railroads. This was true for the wide Pennsylvania Mainline Canal and the narrow Union Canal, the wide Delaware Division canal and the narrow Morris Canal, the different widths on the northern and southern parts of the Miami and Erie Canal, and the larger eastern and smaller western parts of the enlarged Erie Canal in New York.

Nonetheless, there was sufficient uniformity in canal channels and locks so that with a few exceptions canal boats were similar in design and function. Canal travel has been most conspicuously identified with the packet for passenger travel, though most canals were built for the primary purpose of carrying freight. When the alternative was bumpy, dusty, or muddy roads, the apparent ease with which packets glided rapidly and silently over the water attracted passengers to the canals by the thousands, and packet boats were designed to accommodate them.

Boats were sixty to eighty feet long and fourteen feet wide where the locks were ninety by fifteen feet and one hundred feet long when larger canal dimensions were adopted in the 1840s. Packets had a snub-nosed bow and a cabin flush with the hull rising to a long, flat deck, with stairs at either end. Shuttered windows lined the sides, and the craft were often brightly painted. At the stern the steersman sat behind the cabin, beside the great tiller and rudder. The interior was divided into men's and women's sections, with tables for eating in the center and a kitchen in the rear. Along the sides of the cabin three tiers of folding bunks held by chains allowed for sleeping at night and were folded against the sides by day.

Over three decades from the 1820s to the 1850s the lines of these craft were refined and new luxuries offered until, like the later steamboats, they were called "floating palaces." And like the steamboat, clipper ship, and Pullman railway car, they reached fullest development just as they became obsolete. In their peak years they carried one hundred passengers each, ran day and night covering eighty miles in twenty-four hours, and were reported to offer great comfort at minimal cost. The sensation of gliding quietly between close and evenly cut banks was mesmerizing, conveying a sense of effortless speed.

This frequently noted sense of speed seems puzzling when stages traveled more rapidly on open roads. In 1826 Henry B. Stanton wrote: "We shot into Rochester through the aqueduct across the Genesee." And a British traveler wrote of a canal journey in 1834 as he left Schenectady going west: "We shot along, slowly it is true."[46] The latter comment expressed the paradox of a heady sensation of speed even while retaining a sense of reality.

Similar paradoxes are found in almost every aspect of canal travel as recorded in innumerable accounts of the period. A greater number of these describe travel on the Erie Canal because it was part of the Grand Tour to Niagara, and the Erie had the most intensive packet travel for the longest period of time. Although each state canal system developed unique boats and patterns of operation, such accounts record much that was typical on all canals. Moreover, wherever canal travel was undertaken, there were variations by season, boat, degree of crowdedness, and the temperament or social class of the traveler. Thus a resident of Canandaigua, New York, sent word of the impending journey of a friend to Lockport in 1829: "As there will be a canal boat running, she will be able to go with great comfort." By contrast, Francis Granger wrote to Thurlow Weed in Albany in 1832, "I reached home last evening after all the horrors of Canal Boat travelling."[47]

Since the cabin of a canal boat joined the hull in an almost unbroken line, passengers were confined to its long interior, and the roof served as an upper deck. In the heat of summer the cabin of a crowded boat could become hot and uncomfortable. Windows were shuttered but without screens, and flies and mosquitoes tormented the passengers. "Voracious mosquitos" disturbed sleep at night and brought nearly universal complaint.

In fair weather travelers rode on the upper deck, usually reached by a stairway at the stern. On the upper deck the sun beat down, and passengers were frequently warned to "mind the low bridge." A young New Jersey passenger en route to Ohio on the Erie Canal wrote that "if we get our eyes fixed and gazing with delight on anything perhaps at that moment we are loudly called to beware, the bridge, which fright scatters all our pleasures far and wide."[48] Bridges connecting farmers' fields or town streets were built so low (or made so low by high water) that passengers had to prostrate themselves or be swept from the deck. A Canadian riding an Erie Canal packet in 1826 complained that "people are obliged to lie flat upon the deck, (often not too clean) or get down below—got once knocked down by not observing one in sufficient time to take care—rec'd a severe blow, which stunned me a good deal."[49] Some, less fortunate, received broken limbs from striking a bridge. The deck of a canal boat might be crowded with passengers or covered with baggage, but there often was room to move about. A Bostonian going west on the Erie Canal in 1846 wrote to a correspondent, "Everything on the Canal is life and motion. . . . A packet has just passed filled with passengers and a man playing the viol and Gentlemen and Ladies dancing on Deck."[50]

The interior of the long packet boat cabin was often lined with cushioned benches under the windows, and in the center were long tables for eating. At night one end of the cabin was separated for the women by a screen or a painted drop curtain. The benches folded out into beds, and above them two tiers of sacking-bottomed frames for sleeping were suspended from the ceiling by cords or chains, often precarious perches.

In one of his sketches from memory, entitled "The Canal-Boat," Nathaniel Hawthorne wrote of taking one of the "light packets" on the Erie Canal in 1830, boarding it thirty miles below Utica and leaving it before it reached Syracuse. His account was more than a response to the ambiguities of canal travel. He described "this simple and mighty conception," which took the scattered spots of nature and gave them importance. The water of the canal, he wrote, "must be the most fertilizing of all fluids; for it causes towns—with their masses of brick and stone, their churches and theatres, their business and hubbub, their luxury and refinement, their gay dames and polished citizens—to spring up" until one day the canal might pass "between two continuous lines of buildings" from Albany to Buffalo. He passed a boat manned by three Indians who "floated along the current of [the white man's] enterprise." On deck at night with the boat's lanterns lit fore and aft, he thought "that there seemed to be no world, except the little space on which our lanterns glimmered. Yet it was an impressive scene."[51]

But there Hawthorne appeared to reach the limit of his acceptance. The packet was "a dirty canal-boat"; the seventy-mile "long level," on which the canal did not require a single lock, was a "dead flat." There was an "adventurous navigation of an interminable mud-puddle." Moving by "an imperceptible current," the canal made its "drowsy way through all the dismal swamps and unimpressive scenery" along its path. A log cottage might be passed, a lock could create a "little mart of trade," and "the next scene might be the dwelling-houses and stores of a thriving village." Canal travel was "so tiresome in reality" that he rebelled against "the intolerable dullness of the scene," and his perceptions were deadened by "an overpowering tedium."

Hawthorne was not so deadened, however, that he could not record his adverse judgment of his fellow passengers. They were "a vulgar and worldly throng," intruding on "the wild Nature of America." An Englishman was taking notes, and Hawthorne described his company as they might be recorded by "the infernal Englishman." Among the passengers were a Virginia schoolmaster, "a Yankee by birth," who was testing a Schenectady freshman on the conjugation of a Greek verb; a Massachusetts farmer "delivering a dogmatic harangue" on politics and the preservation of the Sabbath from the Sunday mails; and a Detroit merchant, "the worshipper of Mammon at noonday." When the Englishman "lifted his eye-glass to inspect a western lady," she promptly "retired deeper into the female part of the cabin."

Hawthorne watched the towrope catch a passenger by the leg, a new passenger struck down by a bridge, and another "being told to leap aboard from the bank, forthwith plunged up to his third waistcoat button in the canal." His nights were made sleepless by the snoring of his fellows, and when he turned in his berth, he "fell like an avalanche on the floor." His brightest moments came when his packet overtook "a vessel that seemed full of mirth and sunshine. It contained a little colony of Swiss, on their way to Michigan, clad in garments of strange fashion and gay colors, scarlet, yellow and bright blue, singing, laughing, and making merry, in odd tones and a babble of outlandish words."

When the towrope snagged on a branch beside the canal near midnight, Hawthorne stepped ashore to examine "the phosphoric light of an old tree," only to have the packet move out and leave him behind.

"They are gone! Heaven be praised!" he shouted, and set out "with the comfortable prospect of a walk to Syracuse."

Charles Dickens took a packet on the Pennsylvania Mainline Canal from Harrisburg to Pittsburgh in 1842, and for so harsh a critic of American life he found much to praise about this part of his journey. His packet had a "row of little tables which extended down both sides of the cabin" and were put together to make a long table in the center for dinner. The fare included "tea, coffee, bread, butter, salmon, shad, liver, potatoes, pickles, ham, chops, black puddings, and sausages." His cabin had the usual folded cots along the sides, "three long tiers of hanging bookshelves, designed apparently for volumes of the small octavo size," and women were put abed "behind the red curtain." There was a stove and a "little bar hard by." All in all, it was an experience which Dickens "heartily enjoyed," lying idly on the deck by day or "gliding on at night, so noiselessly," with "no other sound than the liquid rippling of the water as the boat went on; all these were pure delights."[52]

On the Ohio and Erie Canal, John Quincy Adams described his journey on the *Rob Roy* from Akron to Columbus, when he went west to dedicate a telescope in Cincinnati in 1843:

> This boat is eighty-three feet long, fifteen wide and had . . . about twenty other passengers. It is divided into six compartments, the first in the bow, with two settee beds, for the ladies, separated by a curtain from a parlor bed-chamber, with an iron stove in the centre, and side settees, on which four of us slept, feet to feet; then a bulging stable for . . . horses . . . ; the third compartment, a dining-hall and dormitory for thirty persons; and lastly a kitchen and cooking apparatus. . . . So much humanity crowded into such a compass was a trial such as I had never before experienced, and my heart sank within me when, squeezing into this pillory, I reflected that I am to pass three nights and four days in it.[53]

The following year, Albert C. Koch, a German paleontologist searching for fossil specimens in America, revealed the variety of experiences offered by a long journey by canal. He took four canal boats across New York on the Erie Canal, entering the canal at Schenectady and changing boats at Utica, Syracuse, and Rochester. The first was crowded to capacity, and he was assigned to one of the three-tiered wall bunks, where he slept poorly and was awakened at daybreak when several passengers began to talk. His second boat was much larger, "more elegant, and we were served better meals." The third was smaller but less crowded, and the late August day was so chilly that "we had a fire in the stove of the cabin, which we found very pleasant." The last boat took him into Buffalo at six in the morning and apparently was little different than the others for he recorded no comment. The seventy-five-hour trip cost $7.75; the railroad would have covered the distance in forty hours but cost $13.50.[54]

After traveling nine hours by lake steamer, Koch took a canal boat from Cleveland to Portsmouth on the Ohio and Erie Canal. It was probably a line boat because it carried two horses on board, which were changed every six hours with the two on the towpath. He found this craft as favorable as any he had taken in New York. "I was satisfied in every respect with the boat on which I traveled," he wrote, "not only was our cabin very elegant and the meals very good, but the captain as well as his subordinates knew their business. I paid only $6.50 and no tips for the entire journey, which took six days." Along the Licking summit he found "a romantic region," where the hundred-foot high cliffs were similar "to the ones of our beautiful Saxon Switzerland" covered with red cedars a hundred feet high.

Koch returned to the East on the Pennsylvania Mainline Canal, which he entered at Pittsburgh. There he took a packet carrying sixty passengers. He felt fortunate to have a bottom bunk, but "the one who slept right above me fell from his bed. He fortunately came to no harm, but in falling gave me an unpleasant shock, snatched up his bed linen and blanket, and climbed back into his bed." After crossing the Portage Railroad between Johnstown and Hollidaysburg, Koch traveled "two days and two nights . . . on the canal and passed through many charming surroundings" before taking the train from Harrisburg to Philadelphia.

Women were provided separate accommodations for sleeping, behind "the crimson curtain," as Hawthorne put it, and may have suffered from the roughness of American society at the time. They gave as varied responses to canal travel as did the men. Anne Royall, the famous American traveler, thought packets "extremely pleasant" when she traveled on the Erie Canal in 1830: "These packets have accommodations for thirty passengers, and very *civil* captains; the ease with which you slip along and the ever varying scenery, is very pleasing to the traveler. The only annoyance is the scraping of the boat against the locks when [the water] is let in, the sudden rising of the boat causes it to drive from side to side, which often awakes those who are asleep; after the first night, however, one gets used to it."[55]

The same year, Caroline Gilman, an English visitor, took a canal boat at Schenectady, "which being a novelty, we wished to test." The "forty persons . . . crowded into this small space" were too many, but "fortunately for us the company was respectable. Groups were soon formed, and various occupations commenced." One gentleman began to read aloud, and she "listened until tears were making their way to my eyes." Though there was much about her long day on the canal boat that was disagreeable, travel on deck was "pleasant enough." The "bobbing one's head down at the bridges" was "somewhat exciting," and she was "convulsed with laughter" at "how comically every body looks" prostrated on the deck.[56]

But Harriet Martineau, the British traveler-critic, warned ladies against travel by canal in her account of western travel in 1838:

> I would never advise ladies to travel by canal, unless the boats are quite new and clean; or at least, far better kept than any that I saw or heard of on this canal. On fine days it is pleasant enough sitting outside (except for having to duck under the bridges every quarter of an hour, under penalty of having one's head crushed to atoms), and in dark evenings the approach of the boatlights on the water is a pretty sight, but the horrors of night and of wet days more than compensate for all the advantages these vehicles can boast. The heat and noise, the known vicinity of a compressed crowd, lying packed like herrings in a barrel, the bumping against the sides of the locks, and the hissing of water therein like an inundation, starting one from sleep; these things are disagreeable.[57]

Catherine Dickinson of Cincinnati, aunt of Emily Dickinson of Amherst, Massachusetts, wrote an account of her journey to Amherst in 1835 over the Pennsylvania Mainline Canal, just a year after it was completed. She took a line boat from Pittsburgh to Johnstown which was crowded with passengers, including "more than twenty clergymen besides elders a plenty." By the time she reached Philadelphia four and a half days later, she wrote, "my veneration for clergy is all fled." After crossing the Allegheny Portage Railroad, she went directly to "another Canal Boat all in readiness" and, reaching Columbia on the Susquehanna, where she left the canal, she was "heartily glad to quit the Canal Boat." She had been "so much crowded that we had not a breath of air at night." But on the last leg of her journey on the Columbia and Philadelphia Railroad, the cars were drawn by horses, which were unhitched while passengers remained in them to endure a storm of wind, hail, and rain before they could proceed.[58]

Her journey has added interest because her father, in Cincinnati, was associated with Lyman Beecher at Lane Seminary. His daughter Harriet Beecher Stowe left a famous account of a similar canal trip east in 1841. Stowe reported on her night on a canal boat in *Godey's Lady's Book*. "Of all the ways of travelling," she wrote, "the canal boat is the most absolutely prosaic and inglorious." Densely packed women and children were crowded into a "little room about ten feet long and six high." There were moments of "a most refreshing slumber" and times when she heard the pleasant "rippling of the rope in the water," but the jars of striking a lock, the importunities of the chambermaid, and the noise of the children made for little rest in "sleeping on a canal boat."[59] Fanny Kemble, the actress, could write that "I like travelling by the canal boats very much," but she protested the "horrible hen-coop allotted to the female passengers." But there was little ambivalence in the report of Frances Trollope after an Erie Canal trip: "I can hardly imagine any motive of convenience powerful enough to induce me again to imprison myself in a canal boat under ordinary circumstances." As she observed American women settling in for their canal-boat journey, she thought they "look like hedgehogs, with every quill raised."[60]

Women who lived beside the canal and knew individual captains could view packet travel quite differently. Caroline Smith of Akron, Ohio, wrote to her sister-in-law in Lima, New York, and advised her, "If you come alone, should you find the boat 'W. SMITH' in Cleveland, be sure to come on it. . . . The boat is in Cleveland every 5 days." When alternative choices were mentioned in her letters, there was no reference to the horrors of travel for women on canal boats.[61]

Family travel by canal could be agreeable and, for the young, an adventure. William Cooper Howells, father of William Dean Howells, moved his family to Chillicothe in 1834, traveling from Portsmouth on the Ohio and Erie Canal. "The journey was a very interesting one to us," he wrote in his *Recollections*, "particularly on the canal, where we had the boat nearly all to ourselves, and spent twenty-four hours in making the fifty miles, we thought this a speedy mode of travel."[62] When the family moved again in 1849, this time from Hamilton, Ohio, to Dayton, the elder Howells chartered a canal boat for the trip. William D. Howells was only twelve, and he later remembered how "the household goods were piled up in the middle of the boat, and the family had a cabin forward, which seemed immense to the children. They played in it and ran races up and down the long canal-boat roof, where their father and mother sometimes put their chairs and sat to admire the scenery."[63]

Travel accounts make frequent reference to music on the canal boats. A traveler's diary relating a trip on the Erie Canal in 1842 recorded "some fair singers, a Singing Book & Bass viol aboard," and he added, "Sung a Considerable." On the way to see Niagara Falls a packet traveler wrote in 1837, "I walked—read—talked—sung—fiddled—eat [sic] very good meals we had too—and looked upon the scenery of the Mohawk."[64]

Two Erie Canal trips by Philip Hone, president of the Delaware and Hudson Canal Company, reveal the varied experiences one could have, the first after the Erie Canal had been in operation for more than a decade and the second a dozen years later. Hone wrote in his diary in 1835: "The boat was not crowded, the weather was cool and pleasant, the accommodations goods, the captain polite, our fellow-passengers well behaved, and altogether I do not remember to have ever had so pleasant a *ride* on the Canal." But his diary entry for a journey on the Erie Canal in 1847, at a time when most packets were vastly improved, was less sanguine: "This canal-traveling is pleasant enough by day time, but the sleeping is awful. . . . The sleepers are packed away on narrow shelves, fastened to the sides of the boat, like dead pigs in a Cincinnati pork warehouse. We go to bed at nine o'clock, and rise up when we are told in the morning."[65]

One's view of canal travel may have depended on the social class of the traveler and whether he or she was a foreign visitor or native American. Caroline Gilman was surprised by the "perfect good humor of the company" as the boat readied for sleeping on her first journey on the Erie Canal in 1836, and she wrote, "Are the scenes I have witnessed really among the same population which English travellers have described? Am I dreaming, when I find only courtesy among the cultivated and quietness among other classes?"[66] Frederick Julius Gustorf came from Hesse in Germany and traveled the Pennsylvania Mainline Canal to Pittsburgh in 1835. He found his boat comfortable, "considering its narrow beam," and reported that "the table is well laid out and loaded with choice food." But on board he met the sculptor Luigi Persico, who was employed on the stonework of the Capitol in Washington, and he wrote that "most of my traveling companions are storekeepers, lawyers, and a few ministers."[67] In contrast to these European visitors, Mary Pratt, wife of the first president of Granville College, traveled on a canal boat from Cleveland to Newark in 1831 in a small cabin with six women, three of whom had small babies and one a young daughter. "Such an unrefined, illiterate, filthy set," she wrote, "I never saw before and do not desire to see again."[68]

Such varied accounts typified canal travel as it declined under the pressure of the speedier railroad. Even in the face of this decline, a Buffalo editor, noting that railroad competition was forcing canal fares downward, still wrote with unqualified praise for packet travel in 1845:

> The superior cheapness of the packets unquestionably is the principal inducement with some travelers to take that mode of conveyance; but so excellent are the accommodations now offered by the packets, and so infinitely more attractive are they to all who travel merely for pleasure, or who wish to travel in the greatest ease and security, they would do a good business were their prices as high as charged on the railroads. The boats that ply between this city and Rochester and Syracuse and Utica, are comfortable and roomy as the cabin of a steamboat. . . . We can conceive of nothing more agreeable than floating tranquilly through beautiful country, such as the canal route has become, with ability to read, write, or talk as the inclination prompts, and with every thing at command that a good hotel can furnish.[69]

Clearly, generalization about canal travel is nearly impossible.

Boats traveled from four to ten miles an hour, pulled by horses or mules that were changed every fifteen miles. Faster speeds washed the banks and were prohibited, a regulation that usually was enforced. Fare on the Erie Canal was typically three or four cents a mile, including food, which was often highly praised, though immigrants traveled at a penny a mile. Competition drove fares down, and stages did not suffer an eclipse because of a new canal on their routes.

Line boats carried both freight and passengers at lower fares. Freight was stored in the middle of such a boat, with rooms for passengers at the fore and for the crew at the stern. There was a variety of freighters. Scows were square at the ends and often open-decked. Deck scows were covered boats and offered more protection to bulk. Lake boats could be towed in rafts. Bullheads had rounded bows and a cabin-like cover that provided dry storage for flour and grain. These freighters carried from seventy-five to ninety tons, often drew only a foot of water, and increasingly dominated canal traffic. Of the 4,191 boats in service on the Erie Canal in 1847, only 62 were packets and 621 were line boats. In the 1840s on the Erie Canal some thirty-seven transportation companies were in business. The major transportation company on the Pennsylvania Mainline Canal was the Western Transportation Company of David Leech, which after 1836 carried both passengers and freight.[70]

Canals were operated by states or private owners and varied over time, yet some elements were common. Down traffic was given right-of-way over up-canal traffic, and packets took precedence at the locks. A boat traveling from a higher to a lower level entered the lock as soon as its forward gates had been swung closed. The lock gates behind the boat were then closed and water was let out of the lock through sluices. The boat was lowered rapidly, the forward gates were then opened, and the boat was drawn out at the lower level; this operation was less easily done than might appear considering the scant three inches open on either side of such long, narrow craft. When one boat passed another, the team of the first boat was checked, the craft glided to the berm bank side, and its hundred-foot towrope fell slack in the water and across the towpath. The overtaking boat then passed over the rope and on ahead. Hundreds of bills of lading and clearances have been preserved, allowing us to follow journey after journey by canal, including one in 1833 for the boat *Equity* of Rochester on the Erie Canal in 1833, which recorded eight tons of furniture from Albany to Buffalo, and noted passengers added and leaving and the interesting item "30 Dutchmen Emigrating."[71]

The number of people employed in operating the canals must have varied widely from the lightly used New England canals and the localistic lateral canals in New York, Pennsylvania, and Ohio to the heavily traveled trunk lines such as the Erie Canal. In the 1840s in New York more than ten thousand people were employed in internal navigation.[72] Among them were the canal boys, who constituted a small army of youths at work on the canals, chiefly serving as drivers for the horses and mules on the towpaths. Included in their ranks were James A. Garfield, who walked (and occasionally rode) as a driver on the Ohio and Pennsylvania canals in 1845 at $14 a month, and Michael Moran in New York, whose career followed the Horatio Alger tradition. Moran's family had come from Ireland to Frankfort on the Erie Canal. His father worked pointing up the locks, and Michael worked as a driver for fifty cents a day. He then became a steersman, bought a canal boat, and became a captain. By 1860 he owned a fleet of canal boats, sending tows down the Hudson River to New York. Ultimately, this led to the fleet of Moran tugs which has dominated towing in New York Harbor.[73]

On both state-operated and privately owned canals, any man could operate any craft that would float, creating an egalitarian pressure against combination or monopoly. Combinations were attempted to standardize fares and freight charges, but such efforts met constant resistance. Tolls were low enough to limit possibilities for profit, and forwarders charged rates that varied by canal and product. Freight charges varied from half a cent to two cents per ton-mile.[74]

The boatmen and their families were a floating population moving across the canals with the opening of navigation in April until November in the North, and many wintered on board their boats. They constituted a new element in the transportation labor force, like the teamsters, rivermen, and steamboat captains. Serving under the canal boat captains were steersmen, drivers, cooks, and chambermaids. They have long appeared in canal fiction and folklore, but few monographs on American canals have examined the men and women who worked on the canals. Only recently have they been given more attention in the new social history.

Like conflicting reports of canal travel, evidence about those who worked on canal boats is elusive and contradictory. They were transients in a society newly drawn to these waterways, and they worked in an environment that was on the edge of social change. Moreover, from the days of the highwaymen or rivermen in the tradition of Mike Fink, they joined an itinerant occupation as a breed apart. Their work was robust, conditioned by capricious changes in the weather or the inexorable progress of the seasons, demanding personal tests of competence or combativeness but bound in some degree by increasing bureaucratic control by state officers or company officials.

Whether on a state or company canal, operations became more systematic, though American canals operated with less bureaucracy than they needed for efficient operation. Schedules of tolls were published; collectors were stationed at canal ports; lock tenders worked around the clock; weighlocks measured cargoes; and clearances and bills of lading were remarkably precise and detailed. Reports of state comptrollers, especially those on the Erie Canal, record violations and fines, down to a fine of $25 in 1834 for four barrels of nuts "not cleared or reported" and a similar fine in 1837 against a captain for seventeen boxes of herring, eight boxes of raisins, and four of codfish "secreted under his stern deck."[75] The New York comptroller was advised in 1833 that "every expedient is resorted to get rid of paying tolls," and with produce and passengers loaded and unloaded at numerous ports on long canals that could not be policed, enforcement of tolls or any other regulation was difficult. That so many tolls were collected, fraud was exposed and punished, limits on speed were largely enforced, and rules of the canal were more observed than ignored seems attested to by the sheer volume of canal trade accommodated and reported.

In a recent study of the Erie Canal, Roger E. Carp has noted the many reports of larceny and fraud on the canal, crews who pilfered cargoes, shippers who sought to avoid tolls, and the frequent drunkenness, theft, and violence aboard canal boats, particularly as perceived by those who sought reform. Lionel D. Wyld has studied the folklore of the Erie Canal and found the fighting canaller a convention in the Erie Canal novel.[76] Along with the riots among the Irish laborers in the 1830s, these elements were part of the harsher side of American life, whether in the new factories of Lowell, Massachusetts, the mobs who confronted abolitionists, or the violence and fighting characteristic of Jacksonian America.[77]

The canals brought new ethnic groups to areas where they were unwelcome. In Mercer County, Ohio, an antiblack mob confronted the freed slave community that came by canal following the manumission of the slaves of John Randolph of Virginia. In 1846, these 383 people were brought to the Ohio River and Cincinnati, where they entered the Miami and Erie Canal en route to lands near New Bremen. The Mercer County mob refused to allow the former slaves to disembark, and they were sent back down the canal to Piqua, where they were dispersed.[78]

Against such accounts of violence and conflict, there were many positive references to the operation of the canals. Praise for the captains and crews, who were welcomed and respected on the canals, was frequent. Letters of recommendation for canal boat captains are to be found such as that in 1844 commending an Erie Canal captain as "a young man of integrity & worthy of trust." When sixteen irate passengers on an Erie Canal packet in 1853 protested that a captain had allowed card playing on board, another group of twenty-four signed a notice testifying to his good sense and noting that "we regard him a gentleman, who, from his urbanity and correct deportment [is] well qualified for the important and responsible station he occupies as Captain of a Packet Boat." And a Lockport, New York, paper noted the opening of the navigation season in

1845: "Capt. Bromley is on the canal of course; he would be very much missed if he was not. A noisy whig as he is, he makes a first rate Packet Captain. The Packets should do a good summer's business; for their owners and managers are indefatigable in deserving business." Far to the west in Indiana in 1852, the *Fort Wayne Sentinel* commended Captain George D. David of the *Caspian*, "who is one of the best and most comfortable fellows living." The British author James S. Buckingham wrote in 1840 on a trip leaving Columbus, Ohio, that "the Captain and his wife who superintended the ladies' cabin, were both very desirious to gratify our wishes, and made us as comfortable as their arrangements would admit."[79]

The career of James K. Moorhead on the Pennsylvania Mainline Canal started with his work as a canal laborer, after which he became a contractor and then operated a packet line based at Huntington on the western part of the canal. Later he became a Pittsburgh industrialist, a Civil War general, and finally a congressman working for the improvement of the navigation of the Ohio River. Moorhead's career was no doubt exceptional, but it suggests the infinite variety of men who worked on the canals.[80]

Violence has probably been overemphasized in most canal literature. There is ample evidence of rules or laws against such conduct, and most travelers' accounts do not report encounters with it. Paul Johnson's study of Rochester on the Erie Canal gives examples of the canal violence deplored by the Rochester gentry. But Johnson also notes that Jonathan Childs, a prominent member of that gentry and owner of the Pilot Line, "the largest fleet of packets on the canal, . . . maintained a strict moral police over his boat crews."[81] In 1826 *Niles' Weekly Register* reported an incident on the Erie Canal: "Erastus Bearcup, a steersman, has been fined and imprisoned in Rochester, on a conviction of using insulting and indecorous language to ladies on another boat, as he was passing through."[82] Reports in the press suggest public censure of fighting between canal boat crews, as when two lines battled for control of packet travel between Buffalo and Rochester on the Erie Canal in 1836 and hired "bullies" who joined in violent affrays wherever their boats met. A Buffalo paper advised, "They have forfeited the good opinion and patronage of the public, and the effectual remedy is to withhold from both any patronage." Such conduct apparently made them exceptions to the norm. The advertisement of an Ohio packet line in 1859 that "each packet also has an iron safe for articles of value" suggests respect for property as much as fear of losses by theft.[83]

Yet the canals hold a conspicuous place in the spirit of reform that spread across the nation between 1820 and 1860. Whitney Cross considers the Erie Canal central to the reforms of the "Burned-Over District" in western New York because it brought New England ideas to the opening of new lands and carried a pluralistic culture to the frontier of the Old Northwest. For example, Charles G. Finney, the great revivalist leader, came west to Rochester by canal. The reform movements begun by the Shakers at Watervliet, the Perfectionists at Oneida Community, the Millerites and the Fox sisters at Rochester, and the Mormons at Palmyra all made the canal corridor in New York a seedbed for reform.[84]

N. Gordon Thomas has described the way the canals carried such reforms to the Midwest, when Ohio became the western center of the Millerite movement. For the Millerite leaders, the rivers and canals of Ohio made cities such as Cleveland, Dayton, and Cincinnati accessible to itinerant lecturers. One of these was Charles Fitch, a follower of William Miller in his belief in the Second Advent and the millennial faith. In the winter of 1844, with the second coming close at hand, Fitch ministered for the new faith in Cleveland. He built a temple with a skylight on top for the journey of the believers to heaven and, in spite of a winter snowstorm, baptized a dozen people in the Ohio and Erie Canal. With other Millerite ministers, Fitch set up his "great tent" in the canal cities of Cincinnati and Dayton, established Millerite weeklies, and went up the Muskingum Valley as well. Miller joined his followers at Cleveland in July 1844 and gave lectures there in July and August. As the Millerite leaders had anticipated, a hundred Adventists from Akron chartered a canal boat to attend these meetings in Cleveland. Miller accompanied them on their return and, "following his usual custom," preached to his audience on board the boat. Leaving Akron, he moved on down the canal to the Muskingum Valley and on to Harrisburg, Pennsylvania, his journey an example of the way canals were part of the spread of the reform movements in the period.[85]

The canals also contributed to the educational impulse in the period. Amos Eaton, a professor at the Rensselaer School in Troy, New York, operated a traveling school of science on the Erie Canal in 1826. One of his students was Asa Fitch, who became the first state entomologist in 1854. Fitch went on board a canal boat and heard Eaton lecture on geology. He also met Constantine Rafinesque, the well-known naturalist, coming east from Kentucky. Returning home, Fitch wrote in his diary, "I was gone only seven weeks and yet how much I have seen! How far I have been! What new ideas I have received! and how greatly my mind has been improved."[86] There were traveling libraries, museums, waxworks, and bookstores on the canals, and on the Wabash and Erie Canal the Spalding and Rodgers Circus Company ran a circus boat, a "Floating Palace," up and down the canal. A handbill of April 1853 announced two performances daily at successive stops at Mt. Carmel, Vincennes, Russelville, Hudsonville, Darwin, and Terre Haute. The price of admission for a seat in the "gallery for Colored Persons" was the same as that for the "Dress Circle, all armed Chairs." The canal boat cabin offered a ready forum for political debate. "When Henry Clay came along on his way to Washington," wrote Richard Smith Eliot of a trip on the Pennsylvania Mainline Canal, "what a chance for the village orator to speak at him and all of us to hear him in response as we sailed from one set of locks to another!"[87]

The canal itself came under the scrutiny of reformers, especially in New York, where the success of the Erie Canal was so dramatic. In a larger cultural sense this was part of the concern for the preservation of American virtue and morality while accommodating the new technologies of the machine, the factory, and the speed made possible by new forms of transportation, as studied by John F. Kasson, though his work has more on response to the railroad than to the canal.[88]

When canal travel was associated with the dreaded spread of Asiatic cholera, it could arouse fear and panic. The disease first appeared in the United States in 1832 and scourged the nation until 1834. It returned in 1848 and every year until 1854, when it virtually disappeared for another decade. These years were the height of the Canal Era, and the disease traveled with trade and transportation. Although cholera was not limited to the waterways, it came with immigrants and appeared in places of filth, sewage, garbage, manure, or contaminated water. Such places were often found on canals. In the first cholera epidemic of 1832, New York City was hardest hit, and the Erie Canal was the major path of immigrants going west. In spite of efforts to quarantine those who were infected, cholera appeared in each successive town on the Erie Canal and soon spread throughout the country.

As Charles E. Rosenberg has written, the terrible disease produced a moralistic response, which attributed the scourge to intemperance, vice, and sin. Cholera was visited especially upon the poor and, in particular, Irish immigrants. Most who witnessed its spread, considered it "primarily a moral dilemma." The response was similar during the second great cholera epidemic of 1849, which was seen as retributive judgment against sin, vice, and disreputable behavior. This second epidemic was most severe in the new cities of the West, in the Ohio Valley, and along the Great Lakes. As cities invoked quarantines and moved toward public health reform, their fears and censure were often directed against those who traveled on the canals.[89]

There was particular concern about the conditions of labor for youthful canal drivers and the moral behavior of the boatmen. Among the national reform movements, temperance and Sabbatarianism were perhaps most conspicuous. The American Bethel Society, founded in Buffalo in 1837, made a major effort to improve the quality of life on the canals. Contributions were collected to assist the men and boys on the canal. Bethel stations were established, and chaplains went out to work among the boatmen. One deacon, M. Eaton, who called himself a "missionary among the watermen," ministered from boat to boat on the Erie Canal in the 1840s.[90]

Eaton was especially moved by the condition of the canal drivers, who numbered some five thousand in 1845. "The wretched condition of these boys cannot be exaggerated," observed a Buffalo petition for assistance to them in 1846, and in 1853 a letter to a Syracuse newspaper urged that city to provide winter labor and schooling "for these poor outcasts." Without doubt, their lives were hard, many were runaways, and they were often abused by captains and disturbed the peace in towns through which they passed. Yet the editor of a Rochester paper in 1845 sought to diminish their reputation as a proscribed class. Although the canal boys were "proverbially a hard set," he wrote, they were "not guilty of half the iniquity they were charged with," and "amid all their roughness and recklessness, they possess many noble traits of character."[91] The Bethel movement spread from New York to other canal states and was part of a larger movement to assist those who worked on the water. In Pennsylvania the United Brethren appointed a canal boat missionary in 1839, and a Lutheran home missionary held services on canal boats in 1848.

The Boatmen's Friend Society was organized in 1830 in New York to promote the "moral and religious improvement" of canal men of all ranks. It resolved to hire a chaplain and sought the cooperation of forwarders and merchants to improve the conditions that were making the Erie Canal a "school of corruption." The boatmen found such charges "rubbish" heaped upon them, and a convention of forwarders responded: "Where, we ask, are the instances of pillage or of any of the abominable vices which this immaculate society has ascribed to the people on the canal?"[92] Reformers continued their efforts by establishing the American Bethel Society in 1837 and the Albany Bethel Union Society, which opened a church in 1843, but the activities of religious reformers declined and were replaced by efforts of boat owners and forwarders to improve the discipline and efficiency of the boatmen.[93]

The movement for temperance reform inevitably touched the canals. The canalside tavern, frequently located at a lock, the drams given to canal laborers, and the drinking by boatmen all made the canals immediate targets. As with all else in the operation of the canals, the response to drunkenness on the canals must be viewed in the context of life at the time. William Rohrabaugh has referred to the nation at that time as the "alcoholic republic." Per capita consumption of alcohol was at an all-time high.[94] Excessive drinking may have been most severe at canal ports at the end of the line such as at Buffalo, where boatmen disembarked after a week-long journey across the state of New York.

Paul E. Johnson has found the temperance question a "middle-class obsession" in Rochester in the late 1820s because the middle class linked drinking with violence as part of the response of employers to new patterns of working-class social life.[95] Boat carpenters and canal crews, like the men who shoveled grain and flour into canal boats, were part of a new working class that threatened the mores of Rochester's shopkeepers. Many of these shopkeepers were wholesalers to country merchants, trading in goods brought by canal from the East or outfitting migrants on their way west. Still, the tensions that led to the temperance movement and other reforms in Rochester were more class-oriented than the result of the rootlessness of new arrivals by canal.

Temperance reform appeared among the boatmen themselves, as when the Erie Canal Temperance Society was organized on a canal boat in 1835 and another was organized on a canal boat in 1843. In 1833 the canal commissioners in New York attempted to enforce a prohibition in canal contracts against the use of ardent spirits of canal laborers, but it proved almost impossible to enforce.[96]

The Sabbatarian movement in the reform era most directly affected the canals because it censured the Sunday traveling of the stages and canal boats alike. Union College was building a new campus at Schenectady just as the Erie Canal was under construction, and its president, Eliphalet Nott, wrote to Governor Clinton in 1826 appealing to him to limit canal travel on the Sabbath. But he admitted that "there is no subject more difficult to treat."[97] The movement against Sunday canal travel was to an extent a catch-all for those who believed that the Erie Canal was so flawed by drinking, fighting, and immorality that it would be called the "Big Ditch of Iniquity." An essay attributed to Lyman Beecher in the *Boston Recorder* in 1828 took the "noble and gigantic work" of the Erie Canal, then being imitated by many other states, as a bellwether for the prohibition of Sunday travel: "If that fearful prostration of the Sabbath, and consequently of morals, which this lofty and sweeping enterprise has carried along with it, cannot be arrested, this example also will become . . . all pervading through the Union."[98]

As early as 1825 and repeatedly thereafter, the New York legislature was petitioned without success to close the locks on Sunday. Mass meetings were held in Rochester and Utica in 1828 to call for boycotts of boats running on Sunday. Governor William C. Bouck, a former canal commissioner, recommended Sunday closing in 1844 and was supported in the New York Assembly by another former commissioner, Horatio Seymour. But they were met with the distinctly modern argument that no law could be passed in advance of public sentiment and that when such sentiment had grown sufficiently, the public would withhold its Sunday patronage from the Erie Canal. Moreover, with the enlargement of the Erie Canal then in progress, how was the canal to reduce its capacity by one-seventh, and what of the interests of the West, which counted on the canal for its markets? The canal remained open on Sunday.[99]

Beginning with the Hudson and Erie Six Day Line in 1827, a succession of canal boat lines on the Erie Canal, among them the Pioneer Six Day Line, the Albany and Michigan Six Day Line, and the Troy and Michigan Lake Boat Line, shut down on the Sabbath. The Pioneer Line was backed by a Rochester Sabbatarian, Josiah Bissell, Jr., and he and his partners announced that they would employ only "men who do not *swear* nor *drink* ardent spirits."[100] Sabbatarian efforts to pressure boat owners and forwarders to stop business on the canal met stubborn resistance, and the Pioneer Line soon failed. But other six-day lines continued to operate on the Erie Canal. In Massachusetts, the Middlesex Canal closed on Sundays for freight boats but allowed packet boats, as long as no "signal horn" was blown. On the Ohio canals some individual captains were known to stop for the Sabbath, but they were conspicuous exceptions to the rule. The Delaware and Hudson Canal, which carried largely freight and few passengers, closed at midnight on Saturdays and opened on Monday mornings. Clergymen traveling on canal boats on the Sabbath frequently conducted services on the packets.

In spite of the censures of reformers, and even the association of the canals with the dreaded cholera epidemics, canals imparted a special flavor to the pathways of travel in the new American nation. For the immigrant arriving at an eastern port, the canal was an inexpensive way to the West, providing travel, food, and housing en route. It was a link to a western steamboat on lake or river. In his classic canal survey written more than sixty years ago, Alvin Harlow included the account of the month-long journey of the Reverend B. W. Chidlaw, who traveled from New York to Cincinnati in 1839, going by steamboat to Albany, Buffalo, Cleveland, Portsmouth, and Cincinnati, all for the scarcely believable fare of $14.75.[101]

For the packet traveler, the canal was a place for social contact, and an egalitarian experience welcomed by some and only endured by others, as passengers ate at long tables and slept on tiered bunks hanging from the sides of the cabin. For captains and crews on packets, line boats, or freighters, the canal meant a life apart as waterborne transients, alternately censured, feared, and treated with affection. In times of tie-ups, delays at locks, tavern brawls, or wintering in clustered boats, they shared a sense of community with others employed on the canals. For the boys who walked behind mules and horses as drivers, the towpath led to a wider world, offered escape for a runaway, and covered miles of monotony, hardship, or cruelty.

Canal life was often characterized by the bustle of basins or wharves, bringing new activity to isolated communities every spring, marked by the distinctive sound of the boatman's horn signaling an approach to a lock. A traveler in Utica in 1840 on the Erie Canal, when the population was only fourteen thousand, wrote of his walk along the canal, where he saw "hundreds of canal-boats, part laden with the produce of the interior ready to start eastward; whilst others, full of foreign merchandise, were bound towards the west." A Buffalo newspaper noted in April 1839 a swarm of canal boats waiting to be off, "covered with trunks, carpet bags, valises and merchandise." William Dean Howells, writing of his boyhood in Hamilton on the Miami and Erie Canal, told how a packet "came in grandly, the deck thronged with people, the three horses in a trot . . . while the steersman put his helm hard aport and the packet rounded to and swam softly and slowly up to her moorings. No steamship arrives from Europe now with such thrilling majesty."[102]

Yet boats passed through wilderness areas, where the yellow light of the canal boat lantern and the almost silent rippling of the boat as it glided over the water added to an overpowering sense of loneliness. This was especially true in western New York in the early years of the Erie Canal, west of the Susquehanna on the Pennsylvania Mainline Canal, and on the canals of western Ohio and Indiana. After a trip in northwestern Ohio in 1851, Moritz Busch left an account of his journey through the Black Swamp on the Miami and Erie Canal. His readers can feel his deep sense of isolation:

> We were still in the Black Swamp, but we were approaching its border. The boat glided slowly in a straight line along its waveless watery path through the moonless forest landscape. The dark woods rose without interruption on both sides. . . . Everything was quiet in the cabin. Not a breath of wind, only the soft lapping of the water on the keel disturbed the deep calm all around. The beam of the hanging lantern below flitted about on the water and ran like an uncertain, nimble will-o'-the wisp through the mists and shadows on the shore. . . . A bell began to tinkle and was answered by ours. Then another canal boat floated past us, with red-curtained, faintly illuminated cabin windows. Then silence and solitude again.

The boat on which he traveled was "overfilled with travelers of all sorts." A dozen others "had to spend the cold, damp night upon the roof of the cabin."[103]

Canals brought new basins, new towns, and new cities. They were part of a new society in which, as Harriet Martineau observed, "everything is of today."[104] The canals extended, with much similarity, over a network of thousands of miles. Not surprisingly, authors seeking to convey the character of life in the young American republic often selected a random canal boat scene, or the dramatic five-lock combine surmounting the Niagara escarpment at Lockport, New York, to illustrate their text. In many ways the canals were a symbol for their age.

7

The Canal Era in Politics and Economic Development

Canals and other internal improvements were integral parts of what is often studied as "political economy." They emerged out of the mercantilist tradition of government encouragement to economic development, stimulating agriculture, manufactures, commerce, and urban growth. Like roads and bridges, canals became part of the national infrastructure, part of the social and economic foundation of the new nation. But they were also subject to the political process, whether that process was viewed as the programs of political parties, as shaped by the force of emergent nationalism, or as an elment in the preservation and advancement of republicanism. Describing the completion of the Erie Canal in 1825, Page Smith called it "the perfect democratic artifact."[1]

If the preservation of republicanism was a primary concern in the political economy of the first two decades of the new American nation, it was a concern that was projected well into the Canal Era. In this context, John Larson has written of the need for the addition to studies of internal improvements of a "national dimension" that might "adequately incorporate republican nationalism and constitutional issues, which rendered internal improvements so crucial to the success of the Union in its early decades."[2]

Historical writing about American canals has exhibited many different approaches, beginning with encyclopedic accounts of transportation, popular accounts recalling the romance of towpaths and waterways, engineering studies, and state histories.[3] But in the last few decades the force of conceptualization has been more stringently applied to canals, most significantly by George Rogers Taylor in 1951, when he considered them part of a Transportation Revolution in nineteenth-century America. Taylor described canals in an emerging national economy and the transportation revolution as "an integral element in industrialization and in the process of economic integration."[4] Canal studies flourished in the 1950s and 1960s, producing books on almost every canal in the United States.

These studies describe the American canals as products of public enterprise, part of what Carter Goodrich has called the "activist state."[5] In spite of the oft-quoted rhetoric of laissez-faire in Jeffersonian and Jacksonian America, the belief that government which governed least governed best, American canals were part of the application of government to economic development. They were, if not the work of the national government, that of local, city, and state governments. And the role of the national government was limited more by state and sectional rivalries than by Jeffersonian or Jacksonian constitutional scruples. There was little disagreement between Federalists and Democratic Republicans on the need for internal improvements, and Republicans were particularly strong advocates of public assistance to internal improvements to preserve republicanism in an expanding nation.

Experience on the pioneer canals of the 1790s convinced most canal advocates that the tasks of long-distance canal building were too large for a company or a corporation, and they also feared monopolistic control. The New York canal commissioners warned against private enterprise in 1811 when they proposed their grand project of the Erie Canal: "Too great a national interest is at stake. It must not become the subject of a job, or a fund for speculation." Yet the emerging corporation and private entrepreneurial leadership in public projects would make many of the canals a mixed enterprise.[6]

It is difficult to explain the transition from Gallatin's plan for a national system of roads and canals in 1808 to Madison's 1817 veto of the Bonus Bill, which would have used the bonus and dividends from the Second Bank of the United States to aid internal improvements. Whether based on Madison's affirmed constitutional grounds or on his underlying fear of corruption and logrolling in the competition for national funds, Madison's proscriptive view of national internal improvement set a precedent for the Canal Era. It was continued with James Monroe's veto of the Cumberland Road Bill in 1822, Andrew Jackson's veto of the Maysville Road Bill in 1830, and James K. Polk's veto of the Rivers and Harbors Bill in 1846. No simple explanation for such denial of federal support for internal improvements can suffice, and such denial was never complete. But as Harry N. Scheiber has noted, ideology had an enduring effect, along with sectional conflict, in sustaining the opposition to federal aid to canals.[7] Thus when the Appalachian barrier to the West was surmounted by canals, they would be built largely by the states.

The development of state programs for canal building was fraught with paradox. Robert Wiebe has called this period the era of state power, characterized by the espousal of comprehensive programs, yet at the same time there was crucial if limited national assistance to canal building. Moreover, the era of state power is also recognized as a time when there were vigorous expressions of nationalism.

Wiebe has written that in this time of emergent nationalism "the national government appeared helpless," while state politics were vibrant. Indeed, in the Missouri crisis of 1819-20, "the real danger to the Union was apathy."[8] The dispersal of power to the states invited the gentry leadership to articulate comprehensive programs to overcome the centrifugal forces of the times. Preeminent in this regard was the American System of Henry Clay. Clay made roads and canals—internal improvements—a cornerstone of his three-part program, which also included a protective tariff and a national bank. The comprehensive programs of Clay, John Quincy Adams, John C. Calhoun, and John Marshall all were replete with the rhetoric of nationalism. If the power of the national government seemed weak, the pulse of republicanism was strong, and it was evident in many of the projects for American canals.

From the time of Gallatin's report of 1808, the national government exercised a crucial role in planning the canal network and in some cases directly aided state or company projects. Less well known but similar to Gallatin's famous report was John C. Calhoun's report to Congress in 1819 as secretary of war, calling for a national system of internal improvements to be built by army engineers. For national defense, Calhoun proposed an intercoastal waterway that would include a hundred miles of canals linking eastern rivers and bays and would cost $3 million. He proposed an interior canal between Lake Michigan and the Illinois River and canals between the Ohio River and the Great Lakes, all as part of a comprehensive system. "Whether we regard . . . internal improvement in relation to military, civil, or political purposes," wrote Calhoun, "very nearly the same system, in all its parts, is required."[9] When the Chesapeake and Delaware Canal Company called on Calhoun and the War Department for assistance in selecting the route for its canal, Calhoun sent the Board of Engineers, consisting of General Simon Bernard and Colonel Joseph G. Totten, to help in 1823, and they had a decisive influence on determining the path of that waterway.[10] When these same army engineers were called to assist on the Morris Canal the same year, they gave that project their support and added in their report: "Of all the means which human ingenuity has devised for facilitating communications between different parts of a country, canals occupy . . . the highest rank."[11]

In spite of President Monroe's veto of the Cumberland Road Bill in 1822, he signed the General Survey Act of 1824 in which Congress pointed the way to federal government sponsorship of roads and canals. This act led to the formation of the Board of Engineers for Internal Improvements in the War Department. The planning functions of the board have been described by Forest G. Hill, who wrote that in 1825 canal studies dominated its work. Bernard and Totten were joined on the board by John L. Sullivan, and they supervised about twenty-four other engineers until the board was abolished in 1831. Before the General Survey Act was repealed in 1838, army engineers assisted on the Chesapeake and Ohio, Morris, and Delaware and Raritan canals, as well as the canals in Pennsylvania and Indiana, though their work was limited to surveys, plans, and estimates. Congress gave cash subsidies to the Louisville and Portland, Chesapeake and Delaware, and Chesapeake and Ohio canals and land grants beginning in 1827 to the states of Indiana, Illinois, and Ohio. And in spite of the anti–internal improvements tone of the Jackson administration in the Maysville Road Bill veto, Jackson signed internal improvement bills totaling more than $1 million a year.[12]

The evolution of American law opened the way for canal construction, whether national, state, or private, as James Willard Hurst has described most fully.[13] Even if canals were predominantly the work of states or private companies, they were accommodated and stimulated by supportive legal institutions. As Harry N. Scheiber has written, "The entire fabric of American law—including both constitutional doctrine and subnational law, both public and judge-made—was reshaped in response to the demands of the Transportation Revolution." And, in turn, "law also shaped transport development."[14] If constitutional limitations slowed national assistance, states and companies helped expand the canal system in uninhibited rivalry with each other.

Canals aided by national subsidy have been seen as national waterways—Ralph D. Gray titled his study of the Chesapeake and Delaware Canal *The National Waterway*, and Walter Sanderlin titled his study of the Chesapeake and Ohio Canal *The Great National Project*. Builders of state and company canals in the East seeking trans-Appalachian trade also saw their works as means to bind the Union together, to overcome divisions between East and West along the Allegheny backbone, and to preserve American republicanism.

Nowhere was this more conspicuous than on the Erie Canal in New York. In a volume of canal documents published in 1821, Charles G. Haines (De Witt Clinton's private secretary) wrote about the Erie Canal then under construction: "But paramount to all other considerations, is the influence to be anticipated from the Western canal in giving strength and durability to our national confederacy. . . . We must bring the north, and the south, and the east, and the west, nearer each other by the attractions of interest. . . . The Western canal will unite the two most populous and powerful sections of the nation, and form one of the strongest safeguards of the union, that either state or national policy is capable of devising."[15] A plaque was placed on the five-lock combine at Lockport, New York, which read: "These works were built by the enlightened wisdom of the citizens of this republican state."[16]

The week-long celebration of the completion of the Erie Canal in the fall of 1825, an event that struck the national consciousness, was redolent with the rhetoric of nationalism and republicanism. John Seelye has noted that the *Memoir* published by Cadwallader Colden at the time of the celebration was the first book published by the new process of lithography to be issued in America, rivaling the Erie Canal itself in the use of new technology in national communication.[17] In the *Memoir*, Colden described the sounds of the signal guns placed earshot distance apart along the Erie Canal and the Hudson, some of which had been used by Admiral Oliver Hazard Perry at the Battle of Lake Erie. "Who that has American blood in his veins can hear this sound without emotion?" asked Colden. "Who that has the privilege to do it can refrain from exclaiming, I too, am an American citizen; and feel as much pride in being able to make the declaration as an inhabitant of the eternal city felt, in proclaiming that he was a Roman."[18]

The elements of nationalism surrounding the Erie and other canals had roots in a cluster of beliefs prevalent in American society. The hand of God was manifest in providing a path for the canal. Americans expressed a sense of history and mission, though with the paradox suggested by R. W. B. Lewis of "the dedicated absorption with history at a moment when it was being claimed that a new history had begun."[19] New Yorkers repeatedly boasted that their canal was an achievement of republican institutions, superior to anything comparable in the monarchical nations of Europe and demonstrating distinctions similar to those articulated in the Monroe Doctrine of 1823. And on the North American continent, the Erie Canal was a nationalistic victory over the constant rivalry of Montreal and Canada for the outlet of the trade of the Great Lakes. As John Seelye has put it, the Erie Canal was "a technological mechanism designed to carry out the geopolitical function of the Constitution—the great Republican machine—which was to assist in the spread of the Union while insuring its stability."[20]

However pervasive such nationalistic beliefs may have been, the application of republicanism to canals made them subject to the political process with its contradictions and vagaries of personal ambition, partisan alliances, and conflicts over means and ends, or as Otto von Bismarck phrased it, the manner in which "laws and sausages are made." At the level of national politics, Henry Clay carried his advocacy of roads and canals into his alliance with John Quincy Adams. Although Adams had questioned the case for the Chesapeake and Ohio Canal in 1807, as president he placed his hopes in a "great national system" of roads and canals. And John C. Calhoun, the advocate of a comprehensive system of internal improvements as Monroe's secretary of war, became as Adams's vice-president "the sable genius of the South," blocking the New Englander's grand plans for a national system of roads and canals. Yet a thread of continuity persisted: the congressional land grants for canals in 1827 and the national subscription to the stock of the Chesapeake and Ohio Canal Company were to some degree the fruits of Clay's American System and Adams's program.[21]

With the national government reduced to a secondary status, states, through a policy of mercantilism, made canals into instruments of economic development. Harry N. Scheiber has provided a cohesive case history of state government in Ohio and its singularly effective program for developing its economy by transportation policy. In a new preface to the 1987 printing of his study of the Canal Era in Ohio, Scheiber described his effort "to systematically relate public enterprise, law, and institutional change to the economy."[22] His study includes the regions, counties, and communities where the state controlled the transportation system and where localism ultimately challenged state planning. In no other canal monograph is the impact of canals on development so clearly evident.

Encouraged by Governor Ethan Allen Brown, the Ohio legislature appointed a canal commission, which decided that a state canal program was indispensable and then set out to use the most productive political means to create it. This required professional engineering surveys and a balancing of sectional and local interests. The legislature put the credit of the state behind the canal program and secured financing in Ohio, New York, and abroad. Once construction plans were determined, the major trunk lines were begun. Out of this process came a mercantilistic design for Ohio's canal policy.

The design also required administration and control. In Alfred Kelley and Micajah T. Williams, as well as other commissioners, Ohio had men of integrity with knowledge of the state who set the patterns for construction, established their own relationships with the legislature, cemented ties to banking institutions, and won sustained public support. The initial result was the completion of the Ohio and Erie Canal and the Miami Canal, both constructed in remarkably successful fashion in 1832. In the broadest sense these canals were the product of the political institutions of the state, the will of an elected legislature, the compromise of regional and local interests, the skill of the engineers and the muscle of canal laborers, and especially the dedication of Kelley and Williams. Though built by the state, these were also canals for the nation. The commissioners noted in their report of 1832-33 to the legislature that these canals would "serve to bind together, by the strong ties of interest, different parts of our state, and of the nation."[23] Ohio's mercantilist policy linked Lake Erie and the Ohio River by canals, and other states were imbued with a similar mercantilistic spirit and launched their own canal projects.

These projects reached their zenith in the mid-1830s, just before the Panic of 1837. Scheiber identified three compelling pressures that almost irresistibly forced an expansion of the public works: egalitarian ideas, the claims to equal benefits from public enterprise, and the widespread belief in the "indirect benefits" related private interests would gain from lowered transportation costs. These were "commonwealth interests," which were used to justify the initial state canal program and were appealed to for further expansion of the Ohio canals. The force of such pressures produced four new state canal projects, financed by the Loan Law of 1837: the Miami Extension Canal, the Muskingum Improvement, the Walhonding Canal, and the Hocking Valley Canal, plus state assistance to private transportation projects. What had begun as a carefully conceived, administratively controlled system of canals was expanded by egalitarian principles, which also reflected "the irrational public mania for internal improvements of the day."[24]

By 1840 the mercantilistic model produced most of what the state's economic planners had worked for, especially on the Ohio and Erie Canal, though the results were more uneven in the Scioto Valley and along the Miami Canal. Farm income rose, the market for manufactures in Ohio grew, and urban development followed. Interaction between agricultural growth and urban commerce "pushed the canal counties ahead in the race for economic development."[25]

Scheiber identified a second phase in America's transportation revolution, following steamboat development and the opening of the Erie Canal. Ohio participated in this second phase as tonnage on western canals rose dramatically and freight rates fell because of competition among the water routes to eastern markets. These developments exercised a dramatic impact on economic growth even before the full transition to the railroad age.

The mercantilist hand of the state was still evident. The state shifted the structure of freight rates to change the patterns of trade and assisted the development of the railroads. Ohio moved toward a modern industrial economy, which was, as Scheiber wrote, the legacy of the Canal Era. His account of the interrelated elements in the development of Ohio's transportation system demonstrates that it was neither accidental nor the working of a laissez-faire economy. It was the consequence of state enterprise, supported by legal institutions and using the political process, to achieve mercantilistic goals.

In the period after the Loan Law of 1837, however, localism moved ahead of state planning. The railroads, which had been assisted by the state, spread rapidly and drove traffic from the canals. The entrepreneurial leadership that had combined private interest and public service in men such as Kelley and Williams was absent from the canal board of the 1840s. In the reaction against state enterprise, the concept of commonwealth interests withered. Although many Ohio canals continued in operation into the twentieth century, the state canals passed into private hands in 1861. Nonetheless, the foundations of Ohio's economic development lay in the mercantilistic planning of the Canal Era, and continuities from that innovative experience remained in banking, in the work of canal engineers who moved into railroads, and in the emergence after the Civil War of the regulatory state. If less strikingly evident, elements of the conceptual pattern of mercantilistic planning and development that Scheiber demonstrated for Ohio can also be found in the canals of other states.

If the mercantilistic goals of economic development by state canal construction or assistance to private canal companies were to be realized, there remained the formidable task of raising the requisite funds. Canals involved high fixed costs, $20,000 to $30,000 a mile, and for most canals construction needed to be supported over a period of several years. After initial funds were secured to begin canal construction, any difficulty or delay in securing further funds meant an irregular rate of progress, abandonment of contracts, work stoppage, and added costs in resuming work.

In addition to small investors, American bankers and men of wealth in the eastern financial centers were the first source of funds. They included John Jacob Astor, William G. Bucknor, LeRoy, Bayard & Company, the Phoenix Bank, and Prime, Ward, and King in New York and the Bank of the United States in Philadelphia. Many of these provided access to the banking houses of Europe, and foreign capital furnished the great bulk of loans that built American canals. European capital could be reached through the Barings in London, for whom Prime, Ward, and King served as agents in the United States and who employed Thomas W. Ward of Boston for advice on American loans. Other European banking houses that furnished capital for American canals were the Rothschilds, Overend, Gurney and Company, Frederick Huth and Company, Palmer, McKillip, Dent and Company, and Hope and Company of Amsterdam.[26]

The canal-building states were fortunate that British capital was available in the period before the Panic of 1837. Englishmen who sought new opportunities for investment abroad were familiar with canals and had witnessed the profitability of their own inland waterways. They had confidence in American state governments and could earn a higher interest rate on loans in the United States than at home.[27]

The New York system by which acting commissioners directed canal construction, while commissioners of the Canal Fund secured funds and managed canal finances, was copied for most state canal projects. Much depended on the personal integrity of the canal fund commissioners, and it was their decisions that either allowed work on the state canals to proceed or set the stage for failure. Following the New York precedent, the states determined that their canals would be built on the credit of the state, which allowed them to sell bonds at 5 or 6 percent to eastern banks and to banking houses abroad. When canals could not be adequately financed by borrowing funds, the states resorted to direct taxation.

The Erie Canal in New York demonstrated the success of that state's investment, and it also showed that sufficient money could be raised by the sale of bonds.[28] The establishment of the New York Canal Fund produced confidence in lenders because it provided the means of paying interest from duties, a tax on lands within twenty-five miles of the canals, and expected tolls. When the tolls from its use, even before it was completed, more than paid the interest on the loans for its construction, it inspired confidence in British money markets and encouraged further investment in American canals.

On the Erie Canal the first loans between 1817 and 1820 were taken by small investors and bore 6 percent interest. By 1821, however, the Bank for Savings in New York City held almost 30 percent of the canal stock.[29] Wealthier investors such as John Jacob Astor and the firms of LeRoy, Bayard & Company and Prime, Ward, and Sands, took more stock after the canal showed signs of being successful. The final loans in 1825 brought premiums of 8 to 19 percent. Foreign investors loaned large amounts but did not exceed Americans in their holdings until 1829. The loans of $7.4 million required to complete the Erie and Champlain canals were raised without difficulty. Beginning in 1822, loans were taken by the British firm of Thomas Wilson and Company and by Baring Brothers, who bought up earlier loans as well. The latter loans brought premiums of 8 to 10 percent. When the Canal Fund debt was reduced to $3.4 million by 1836, only $548,000 was held by Americans.[30] Nathan Miller has described the Canal Fund as a "development bank" in New York. Its income was sufficient to retire the canal debt rapidly and still use the fund as a source of credit acting through the banks of the state. In 1831 the New York legislature authorized loans from the Canal Fund to the banks. Such loans helped New York banks through the financial stringency of 1834 and assisted the New York City banks in recovering from the great fire of that year.[31]

In 1835, however, the legislature embarked on an expanded canal program, including the lateral canals and the enlargement of the Erie Canal from dimensions of forty feet by four feet to a width of seventy feet and a depth of seven feet, with construction of enlarged double locks. A

Whig administration in 1838 proposed such ambitious plans to implement and expand this program that it was accused of planning to borrow a million dollars for ten years, thus incurring a "Forty Million Debt."[32]

But the Panic of 1837 created such pressures on the financial markets that New York banks suspended specie payments. The Democratic commissioners of the Canal Fund loaned stock to eight banks in New York City to sustain them and enable them to move toward resumption of specie payments. The primary goals of the commissioners were the preservation of the credit of the state, payment of the interest in specie on a canal debt, which in 1837 stood at $6.3 million, and ultimately to retain the support of the English banks. Even though the rest of the state had moved to a paper standard, as Nathan Miller notes, the commissioners maintained specie payments "because of their realization that the American economy, nurtured from London, was dependent on this source of credit to maintain the rate of economic development to which it had become accustomed." New York bankers cooperated with the commissioners because "they had a stake in the credit of the state, intimately related as it was to private credit in an economy that depended on an influx of capital from abroad."[33]

To pursue the expanded canal program authorized in 1835, the state was able to sell $5.4 million in new bonds, but after the Panic of 1837 the state had to curtail its program. The stop and tax law of 1842 halted construction of public works. A tax program that had been urged as soon as the lateral canals were begun was put in place, and the enlargement of the Erie Canal was gradually completed by 1862. By 1851 $16 million had been spent on canal construction since the expanded program was authorized in 1835 and a "Nine Million Bill" was enacted to further construction by the sale of certificates based on future canal revenues. But this act was found to be unconstitutional, and a new measure that increased authorization for an expenditure of $10.5 million on the New York canals was enacted through a constitutional amendment in 1854. This decisive act made the completion of the enlarged Erie Canal mandatory.[34]

The state that came closest to matching New York's achievement in financing its canals was Ohio. De Witt Clinton not only turned the first spadeful of earth on the Ohio canals, but he helped Ohio Governor Ethan Allen Brown and the Ohio canal commissioners obtain $4.5 million in eastern capital.[35] Loans for the Ohio canals were taken in New York by banking houses such as those of John Jacob Astor, William G. Bucknor, and Prime, Ward, and King, which had profited from their investments on the Erie Canal. An initial loan of $400,000 at 5 percent in New York provided for the start of construction in 1825. Ohio banks took some loans in 1826, but $800,000 of the million-dollar issue in 1826 was taken by John Jacob Astor at 6 percent. The following year, the completion of the Ohio and Erie Canal between Akron and Cleveland helped the commissioners to sell $1.2 million in bonds at a premium. Most were taken by a syndicate of Philadelphia bankers and some were taken by Prime, Ward, and King. The same New York and Philadelphia investors loaned $1.2 million more in 1828 and additional amounts in 1830. Ohio began to receive revenue from the sale of canal lands in 1832, and that year the final loan was taken, again at a premium reflecting Ohio's high credit rating.[36]

After 1836, however, Ohio faced financial problems similar to those in other canal-building states. Its public works expansion under the Loan Law of 1837 was similar to programs elsewhere, and progress was hobbled by the same financial stringency following the Panic of 1837. Yet Ohio was able to secure funds to go on with its canals, and as Scheiber has written, its escape from repudiation or default was "something of a miracle."[37] Although Ohio banks suspended specie payments, the new Board of Public Works, which succeeded the Canal Commission in 1836, was able to meet interest requirements in equivalent specie payments. This policy was similar to that adopted in New York, with like advantage to Ohio's credit. Although no new bond issues were needed between 1832 and 1836, in 1837 a half-million-dollar loan in 6 percent bonds was sold to a Prime, Ward, and King syndicate in New York. The next year Prime, Ward, and King took more than $1 million, though skillful management was needed to negotiate the loan, which was secured on problematical terms.

The largest loans were taken by Ohio banks, $2.87 million in the period 1836-39, compared to $2.26 million from New York banks. The Ohio Life Insurance and Trust Company of Cincinnati loaned the greatest amounts, $1 million in 1838 and $665,000 in 1838. This support from Ohio banks often involved desperate, and sometimes illegal, measures. In 1841 the board members signed personal notes to guarantee some of the loans, even at risk of their own fortunes.[38]

At a crucial juncture in 1842 Baring Brothers in London purchased $400,000 in bonds, which allowed the board to make a scheduled interest payment. Taxes provided for in the canal law of 1825 rose each year in the 1840s. A bond issue in 1844 was the last needed for new canal construction, and its sale indicated that Ohio's credit was once again stable. In 1850 the Ohio debt stood at $19.38 million. Construction on the public works and by companies aided under the Loan Law of 1837 had cost Ohio $18.8 million. The "miracle" of Ohio's escape from repudiation or default, as Scheiber has concluded, resulted from the broad public support for its public works, the early provision for direct taxes, the strong banking alliances in Ohio, New York, and London, and especially the expedients sanctioned by the legislature.[39]

Pennsylvania built the largest network of public and privately built canals and accumulated the largest state debt for internal improvements. In contrast to New York and Ohio, the financing of the Pennsylvania canals was fraught with difficulty and the losses the state suffered by the time of their sale in 1857 made them a financial failure.[40] Like New York and Ohio, Pennsylvania built its canals on the credit of the state, and financing was in the hands of commissioners of the Canal Fund. But the tax revenues provided for in the act of 1827 were inadequate, and no protection for the Canal Fund was established as in New York and Ohio, nor was Pennsylvania as fortunate in having skillful credit managers.

Initial loans at 5 percent interest did well between 1826 and 1828. But there were no bids on the loans of 1829, and the canal commissioners could not meet the interest due in 1830. A limited program of taxation was enacted in 1831, and inadequate as it was, in 1834 loans for the public works were taken at a premium. These funds allowed the completion of the Mainline Canal and some of the lateral canals by 1835, when the state debt stood at $24.5 million, $22.4 million of which had been spent for canals and railroads. The following year the distribution of surplus revenues by Congress allotted almost $3 million to Pennsylvania, and the public works received a bonus from the chartering of the United States Bank of Pennsylvania. As part of its charter, the bank was required to provide a bonus to Pennsylvania of $2.5 million and to loan up to $6 million to the state or private companies for internal improvements.[41]

If the bonus and the loan requirements were the price paid by Nicholas Biddle for the charter of the United States Bank of Pennsylvania, the credit of the state became dependent on the stability of the bank. After the Panic of 1837 struck, Pennsylvania was unable to sell the bonds offered in 1839, and the United States Bank was required to take them. With the failure of the bank, Pennsylvania's credit abroad declined. British citizens held two-thirds of the state debt, which totaled $34 million, and they publicized their outrage at the inability of Pennsylvania to pay interest on its loans, which it could not do in 1842, 1843, and 1844. "When Pennsylvania defaulted on its interest payments on August 1, 1842," wrote Ralph W. Hidy in his study of the Baring Company "for the Barings the nadir of the depression had been reached."[42] The American agent for Baring Brothers, Thomas W. Ward, launched a campaign to restore interest payments on loans due the British, especially those in Pennsylvania. After a campaign virtually orchestrated by Baring Brothers, Pennsylvania resorted to a comprehensive tax law in 1844 so broad-based that the state could resume interest payments on its public debt the following year.[43]

Meanwhile, a movement began to sell Pennsylvania's public works, which marked the turning point against state development through internal improvements. Although work continued on the branch canals on the upper Susquehanna and on the enlargement of the Delaware Division Canal, the canals were sold in 1857. The financial loss to Pennsylvania up to the time of their sale has been estimated at $58 million. The constitutional provision for the sale of the Pennsylvania canals also ended the use of the mixed corporation for state investment in private canal companies. By 1857 Pennsylvania had abandoned the principles of public enterprise on which its vast canal system had been built.[44]

For Baring Brothers in London, however, the loans made to Maryland for the Chesapeake and Ohio Canal were even more difficult to accommodate than those of Pennsylvania in the depression years of the early 1840s.[45] The Chesapeake and Ohio Canal Company was assured of public support in 1826 by Congress, Virginia and Maryland, and the District of Columbia. Yet the company suffered from a chronic lack of funds for construction, which led contractors to abandon their contracts and the Irish laborers to riot in the 1830s. National support ceased, and Virginia provided little. The canal finally found most of its support in appropriations and loans from Maryland, which had subscribed to half a million dollars of its stock in 1826. In 1834 Maryland loaned $2 million to the canal company and the state sold 6 percent bonds to cover the loan. Thomas W. Ward recommended the bonds to British investors and they sold at a premium of 17 percent.[46]

But Maryland's $3 million stock issue for the canal company in 1836 faced great difficulty. It was part of an $8 million loan package, which included other internal improvements. Unable to sell these stocks abroad because of the Panic of 1837, the state transferred $2.5 million in bonds to the canal company. The company then sold some in New York and Baltimore, and the Baltimore merchant George Peabody acted for the company to sell the rest in Europe, on which the company realized only 70 percent.[47]

Maryland's public debt for all of its internal improvement projects stood at $15 million in 1840, and the state defaulted on its interest payments until 1848. Like Pennsylvania and New York, Maryland responded initially with a direct tax in 1841. Returns were inadequate, however, and the state attempted to sell all its investments in 1843. This failed, and there was clamor for repudiation. Baring Brothers was the official foreign financial agent for both the canal company and the state. As in Pennsylvania, Thomas W. Ward acted for the Barings to stave off repudiation. The canal company benefited from a compromise agreement in 1843 to settle its debt to Baring while the latter continued its efforts to get Maryland to resume interest payments.[48]

Contrary to its earlier policy, the Baring company sought to influence the election of a Whig governor, Thomas E. Pratt, in 1845. The London company employed John H. B. Latrobe, attorney for the Baltimore and Ohio Railroad, to direct the campaign in Maryland. Latrobe had been engaged by Thomas W. Ward to lobby in Pennsylvania for resumption of payments there. In addition, the Barings financed a campaign led by George Peabody to elect a legislature favorable to resumption of interest payments in Maryland.[49]

Pratt was elected, but resumption did not immediately follow. Maryland tried to complete the Chesapeake and Ohio Canal to Cumberland as a means of gaining revenue and easing the need for taxation. The state allowed the canal company to issue $1.7 million in bonds. After the Barings refused to purchase them, the company sold them in the United States and secured funds for further construction. The canal company, the Barings, Governor Pratt, and George Peabody exerted pressure for a resumption of interest payments on Maryland's debt. Improved economic conditions in the state and a new tax law enabled resumption of payments in 1848. The debt in arrears for the Chesapeake and Ohio Canal and the Baltimore and Ohio Railroad was discharged by 1851.[50]

Virginia's support for the James River and Kanawha Canal was much like that of Maryland for the Chesapeake and Ohio Canal Company. But unlike Maryland, Virginia maintained flawless credit throughout the long years of canal construction. Although state support took changing forms, the James River and Kanawha Canal project faced a constant shortage of funds, which finally thwarted its efforts to complete a water route from Richmond to the Ohio River. After the collapse of the James River Company, the canal came under state control from 1820 to 1835. In the latter year the canal was continued under the newly organized James River and Kanawha Canal Company.[51]

While the canal was under state control, Virginia secured loans on the state's credit which reached $1.23 million by 1826. But then the legislature refused further funds and work stopped. In the chartering of a new company in 1832, Virginia subscribed to three-fifths of its stock, taking new loans of $1.2 million.[52] This stock issue allowed the canal to continue, but in the 1830s the state shifted the major share of its investment to railroads and the canal struggled forward with inadequate funds.

The Virginia state debt reached almost $6 million in 1839, about one-third of which was held in England and France. Although other states found it difficult or impossible to borrow abroad, Virginia's credit remained strong. The American minister to England advised the Virginia legislature in 1839: "This is the time for Virginia to make a loan for the purposes of internal improvements. I can make any loan for the state, to the amount of five or ten millions." The state, however, refused to borrow further for the canal.[53]

Almost half of Virginia's debt before 1831 was held in the state, much of it taken by cities and towns. By 1834, Richmond had subscribed about $1.1 million of the state debt, and in 1837 the city made a loan of corporation stock to the canal company. Under the stress of the Panic of 1837 the Bank of Virginia made additional loans, which raised more than $1 million for further construction. The legislature allowed the canal company to borrow additional funds on its own credit, to be guaranteed by the state. But the misapplication of this power by James Hamilton of South Carolina, who acted for the company in Holland, denied the company foreign support. In 1842, an act not unlike the restrictive laws on canal construction in other states at the time brought the company under stringent financial control by the state. With the limitations imposed by this measure, construction on the canal could not continue beyond its mountainous terminus at Buchanan.[54]

Unlike Maryland and Virginia, New Jersey did not invest in the canals that were expected to aid in the development of the state, but it chartered the Morris Canal and Banking Company, which built the Morris Canal in the northern part of the state to bring Pennsylvania coal to Newark and New York. As a banking company, the Morris company had an indirect relation to the Pennsylvania canals through the investments of Nicholas Biddle and the United States Bank of Pennsylvania, and the company's failure had a disastrously direct effect on the Wabash and Erie Canal in Indiana.

Close relationships between canals and state banks were prevalent throughout the Canal Era. Canal banks were created on the New Haven and Northampton Canal and the Blackstone Canal in New England, and the Delaware and Hudson Canal Company was given banking privileges when it was chartered in New York. New Jersey chartered the Morris Canal and Banking Company to get a canal, but the banking function of the company soon overshadowed its canal management. The company's effective headquarters was in New York rather than New Jersey, and in 1835 Wall Street speculators gained a corner on its stock. The corner was broken, but the episode tarnished the company's banking reputation, which was considerably restored when Louis McLane, former secretary of the treasury, became its president.

McLane gave close attention to the improvement of the Morris Canal and its extension to a terminus on the Hudson River. Yet the canal remained subordinate to the banking functions of the company. This was a reversal of the policy of the earlier president, Cadwallader Colden, whose interest in internal improvements had begun with his association with the Erie Canal. The company's first foreign loan was $750,000 borowed in 1830 from the Dutch banker Willem Willink, Jr., who took a mortgage on the canal. This infusion of funds allowed completion of the canal two years later.[55]

After 1835 the banking activities of the Morris company became increasingly speculative, if temporarily profitable. New Jersey continued to grant the company legislative favors in support of the canal, and the company paid dividends in 1836 and 1837. But as President McLane declared, these profits were accumulated with "but a small contribution from the canal."[56] The company leased the canal to the Little Schuylkill and Susquehanna Railroad, which was planned to reach into the anthracite coal fields of Pennsylvania but had not yet been built. As Earl J. Heydinger has written, this was part of a Morris company "*dream* of a canal-rail line from New York to Buffalo."[57] The high-flying banking activities of the Morris company collapsed in the Panic of 1837, but not before they had contributed to the major share of the losses of the state of Indiana on the bonds of the Wabash and Erie Canal. Through Dr. Isaac Coe, the Indiana canal fund commissioner who became a director of the Morris company, the company purchased Indiana bonds beginning in 1836, and the collapse of the company contributed to the default on the internal improvement debt of that western state.

To build her public works, Indiana authorized initial loans of $600,000 between 1832 and 1835 at 6 percent and then a $10 million loan to finance the Mammoth Act for internal improvements in 1836. A board of three canal fund commissioners, created in 1832, set out to sell state bank bonds, internal improvement bonds, and Wabash and Erie Canal bonds at 5 percent, getting the loans in New York and abroad. They secured loans of $1.5 million for the Wabash and Erie Canal as well as almost $1 million for the state bank in 1834 and 1835 from New York banks and J. J. Cohens and Brothers in Baltimore. But in 1836 Dr. Coe, one of the commissioners, began the ruinous financial relations between Indiana and the Morris Canal and Banking Company. Acting alone and fraudulently, Coe arranged some five loans totaling $1.8 million by 1838 in which the Morris company would market Indiana securities in the United States and abroad, only one-fourth of which the state received. Coe became a stockholder in the Morris company and pocketed a part of the funds. The Cohens company failed in the Panic of 1837 owing Indiana almost $300,000, and when the Morris company failed in 1839 it owed Indiana $2.5 million. Through the failure of these companies and embezzlement by state officials, Indiana lost some $3.5 million. Since the Morris Company was the major source of funds for Indiana's public works, the collapse of the company's banking and canal venture in New Jersey had a devastating impact on the Indiana canals.[58]

Work on the Indiana canals stopped in 1839, and by 1841 the state could no longer pay interest on its debt. All the public works but the Wabash and Erie Canal were transferred to private companies and on that canal work resumed with payment by canal scrip. The legislature authorized 7 percent bonds, which could not be sold, and as in other states there were calls for repudiation or taxation. Land sales brought in $1 million and the canal was opened from Lafayette to Toledo in 1843; another land grant in 1843 helped the state finish the canal to Evansville. But the state debt reached almost $14 million.[59]

Although the record of the early canal commissioners was marred by Coe and the Morris company, Charles Butler attempted to restore Indiana's credit and finish the canal. Butler was born in the Hudson Valley in New York and had moved to western New York to practice law along the path of the Erie Canal. As a western land developer he invested in lands in Toledo, at Huron (present-day Port Huron, Michigan), and Chicago. His investments were built on credit, and he cultivated close relations with eastern capitalists, which helped him negotiate a settlement of the Michigan state debt before tackling that of Indiana.[60]

In 1844 Butler was employed by a group of English bondholders headed by John Horsely Palmer and including the Rothschilds and Hope and Company of Holland to preserve their investments in the Indiana state debt against the threat of repudiation. John Denis Haeger has described Butler as the "epitome of the promoter."[61] Yet his personal integrity, entrepreneurial innovation, and tenacious interest in economic growth and public development made him representative of many of the individuals responsible for the building of American canals. In the Indiana legislature the "Butler bills" of 1846 and 1847 put the Wabash and Erie Canal into the hands of trustees who would operate it and complete it to Evansville. For the investors whom he represented he won a final compromise in which the state would levy a tax and payments to the bondholders would come half from the proceeds of taxation and half from canal revenues. The bondholders agreed to provide an additional $800,000 for the completion of the Wabash and Erie Canal, which would be the security for their investment.[62] The trustees, one of whom was Charles Butler, performed well in an impossible financial situation, and the canal was finished. With tireless skill and partial success, Butler sought to preserve the rights of eastern investors and foreign bondholders against the threat of repudiation.

The feat Butler accomplished in Indiana was attempted earlier in Illinois by his friend and associate Arthur Bronson, who also represented New York and British creditors. At the outset in 1835, Illinois attempted to pay for its canal program, including the Illinois and Michigan Canal, by a loan of $500,000, negotiated on the credit of the state at 5 percent. The state sold bonds to Macalester and Stebbins in New York, who sold them to British investors. Interest was paid from land sales from the congressional land grant of 1837.[63]

When Illinois embarked on its great internal improvements program in 1837, it authorized a loan of $8 million in 6 percent bonds. The state banks were involved in this arrangement as the state subscribed to $3 million of their stock and they were to act as fiscal agents for the Illinois and Michigan Canal. In spite of the Panic of 1837, the legislature authorized a loan of $4 million at 6 percent for the canal. When the stock could not be sold in London, it was sold as favorably as possible in New York to the Phoenix Bank and to the United States Bank of Pennsylvania, but on terms that violated the authorizing act. Even though litigation over this sale hurt the credit of Illinois in London, the fund commissioners made a sale to the John Wright Company in London in 1840, but this company soon failed.[64]

In 1840 most of the Illinois internal improvement projects were abandoned, although work continued on the Illinois and Michigan Canal. Contractors were able to secure for the state a sale of $1 million in stock to Magniac, Jardine and Company in London, and money was borrowed in Philadelphia to pay interest in 1841.[65] But the crisis in Illinois did not allow further interest payments until 1846. Principal contractors such as William Ogden urged the legislature to enact a direct tax, and work stopped in 1841.

The state debt of Illinois amounted to $15 million in 1842. There was talk of repudiation, as in Pennsylvania, Maryland, and Indiana, but there was also confidence that a completed Illinois and Michigan Canal would assure the payments to bondholders. Few men were more concerned about a resolution of the debt crisis than Arthur Bronson, who was a large-scale investor in lands in Chicago and northern Illinois. Bronson came from a family of financiers in New York, and his cautious methods were those, as Haeger notes, of a "conservative speculator" on the frontier. In partnership with Butler, he invested in land at several locations around the Great Lakes. Both men were directly interested in the Illinois and Michigan Canal, and in 1833 they proposed a private company capitalized at $1 million to construct it, a proposal they withdrew when the canal project moved forward under the state. Nevertheless, Bronson became involved in the purchase and sale of state bonds issued for the waterway. His desire to find a solution to the Illinois debt crisis was strengthened by his association with Butler, who represented foreign bondholders directly, and by the position of Bronson's brother-in-law James B. Murray, who had been president of the Morris Canal and Banking Company from 1839 to 1841. Moreover, Bronson had sold Michigan bonds abroad in 1838.[66]

In 1843, three years before Butler used a similar approach to the state debt of Indiana, Bronson prepared a plan for repayment of the debt and completion of the canal in Illinois. His plan, though not immediately implemented in full, was adopted by the Illinois legislature. Illinois would use the canal as security for a new loan of $1.6 million. The canal would be placed in trusteeship, its revenues would go toward the interest on the new loan, and a new tax would be enacted. The latter requirement was resisted in the legislature, but it was insisted upon by the British bondholders whom Bronson represented. Finally, in 1845 the requisite tax was adopted, and in 1849 it was increased, as added protection to the bondholders. With this assurance, Magniac, Jardine and Company, holders of most of the Illinois debt, joined with Baring Brothers to raise the $1.6 million loan for the completion of the canal. The trustees who took over the canal were W. H. Swift of the U.S. Army Topographical Engineers, David Leavitt, and Jacob Fry. Leavitt was president of the American Exchange Bank of New York, and he secured the assent of the New York bondholders. The trustees supervised the completion of the canal by 1848, sold the canal lands, and paid off the loan by 1853.[67] Bronson died in 1844, before his plan could be realized, but he had protected the interest of New York and British investors and allowed resumption of interest payments on the Illinois debt by 1846. Charles Butler continued his investments in Chicago, and that same year he used Bronson's Illinois settlement plan as a model for his mediation of the state debt of Indiana.

As the Barings believed when they contributed to the election of a Whig governor in Maryland in 1845, canal financing was often related directly to the outcome of party contests during the Canal Era. Canals were constantly immersed in the political process, and parties differed on the issue of internal improvements. In the implementation of Clay's American System, the Whigs have traditionally been regarded as stronger advocates of internal improvements, including canals, than the Jacksonians. At the outset of Jackson's presidency, his Maysville Road Bill veto appeared as a reversal of the encouragement of internal improvements by John Quincy Adams that was evident in his assistance to the Chesapeake and Ohio Canal. In fact, during Jackson's administrations, Congress appropriated more than $1 million a year for internal improvements. In 1834 Jackson sounded like the Delphic Oracle: "I am not hostile to internal improvements, and wish to see them extended to every part of the country." But he added, "If they are not commenced in a proper manner, confined to proper projects, and conducted under an authority generally conceded to be rightful . . . a successful prosecution of them cannot be reasonably expected."[68]

In state politics, however, party attitudes on canals were so varied as to weaken the usefulness of any such generalizations. Democratic administrations were often as fully committed to economic development through state canal programs as were the Whigs. Moreover, Democrats split into radical and conservative factions. The Radical Democrats, sometimes called Locofocos or Barnburners because of their views on banking or slavery, tended to oppose the use of state loans for canals; and the Conservative Democrats, sometimes called Hunkers, were inclined to support state canal programs.

Political contests and party fortunes often turned directly on the issue of canals. De Witt Clinton rose to power in New York on his championship of the Erie Canal, and his removal from the post of canal commissioner was decisive in his reelection to the governorship in 1824. William Henry Seward's gubernatorial campaign of 1842 staked all on the issue of the enlargement of the Erie Canal in spite of the Panic of 1837, and Seward went down in defeat. In 1849 Samuel B. Ruggles, who had been a Whig canal commissioner, wrote that the history of the Erie Canal since 1825 "would constitute in good degree, the political history of the state." And in 1852 James Watson Webb, the Whig editor in New York City, warned that "the Enlargement of the Erie Canal is a measure which the Whig party of this State cannot abandon if they would. Their whole history of the last thirty-five years has been pledged to it."[69]

In Democratic party politics in New York, skillful shifts on the issue of canals were a hallmark of Martin Van Buren's political career. First opposing the Erie Canal because of its Clintonian origins, Van Buren gave it crucial support in 1817. He claimed it as his own in 1820 and then turned against national improvements under Jackson partly because New York had built the Erie Canal alone. As Nathan Miller has demonstrated, the Democratic commissioners of the Canal Fund actively used that fund as an instrument of economic development. Yet in the New York legislature, the Radical Democrats led by Michael Hoffman met the crisis over the "Forty Million Debt" alleged to have been incurred by the Whigs with the stop and tax law of 1842, which suspended construction on the public works. This stringent step was paralleled in state after state at about the same time. The Democratic governor elected in 1841, William C. Bouck, however, was a Conservative who had been a popular canal commissioner. He wanted the enlargement of the Erie Canal and construction of the lateral canals to go forward. He was backed by Henry

Seymour, the Democratic chairman of the Committee on Canals, who lived at Lockport beside the Erie Canal. Bouck devised a plan in 1844 for continued canal construction without incurring new debt. "Indeed," wrote the New York historian De Alva S. Alexander regarding the legislative session at which the plan was adopted, "the history of the session may be described as the passage of a single measure by a single man whose success was based on supreme faith in the Erie Canal."[70]

Paradoxically, Michael Hoffman was a Radical Democrat who lived at Herkimer on the Erie Canal in the Mohawk Valley. He was the man most responsible for writing the stop and tax law into the New York constitution of 1846, which also prohibited loaning the state's credit to a mixed corporation. The Mohawk Valley owed its development to the Erie Canal, and perhaps in no other valley in the nation could canal life have been witnessed more closely.

To the historian of Jacksonian politics in New York, Lee Benson, the issue was largely moot. He found ethnic traditions most important in New York voting patterns and challenged the "time-honored but fallacious assumption" that the Erie Canal had the major impact on the state's prosperity. Few observers, however, could deny the importance of canal offices in party patronage. Every shift in political power brought new engineers, superintendents, and lock tenders to the entire line of the Erie Canal.[71]

Partly because of the success of the Erie Canal, the Pennsylvania Mainline Canal was undertaken with the overwhelming support of the electorate, without regard to partisan identification. Pennsylvania politics differed from that in New York, however, because Jacksonian Democrats dominated the state until 1832 and claimed the greatest popular support until the Civil War. Louis Hartz has written that "the state was as ardently wedded to the American System as any in the Union." After the debt crisis of the early 1840s, the decision to turn the Pennsylvania canals over to private ownership cut across party lines. The sale was authorized by a Democratic senate in 1855, supported by a Democratic governor, William E. Bigler, and enacted by a Whig-dominated legislature in 1857.[72]

Party patronage, however, affected canal construction from the beginning. The record of the young engineer James D. Harris reflected the power of political appointment. He was removed from work on the Mainline Western Division in 1829, appointed to the North Branch Canal, and removed from it in 1839 because of his political affiliations.

The administration of the Whig governor Joseph Ritner (1836-38) made blatant political awards in hiring contractors on the North Branch Canal. But his Democratic successor, David R. Porter, was no less partisan in canal administration. Porter's superintendent on the Tioga line of the North Branch Extension announced that he would drive every Whig and Anti-Mason off the line, "estimating them off" through the canal engineers. Harris was identified with the Ritner administration and forced to leave the canal.[73] In the politics of canal management, state policies sometimes emanated from the offices of the Leech Transportation Company with a direct effect on the operation of the Pennsylvania Mainline Canal.

On the Ohio canals both Democrats and Whigs carried out the state's mercantilistic policies. There the broad political support for the canal program rested more on the compromise of sectional interests than on the resolution of political conflict. Micajah T. Williams and Alfred Kelley, who cooperated as architects of the program and served as the first two acting commissioners, differed in party affiliation. Williams became a Jacksonian Democrat and Kelley became a Whig. In the mid-1830s Democrats joined Whigs in support of an expansion of the Ohio canals and supported the loan law of 1837, which allowed state investment in private companies. As Harry N. Scheiber has written, "The two parties vied with one another to take exclusive credit for the expanded program of internal improvements: the Democrats as defender of the public works against predatory Whig financiers, the Whigs as defenders of the state's credit alliance with the bankers and investors who were providing the capital necessary for bond sales."[74]

Under the Jacksonian governor Robert Lucas, the Democratic legislature in 1836 replaced the commissioners with a Board of Public Works composed entirely of Democrats. As party patronage became more prevalent, engineers were removed, toll collectors changed, and canal policies came more closely under legislative supervision. Not surprisingly, however, when popular confidence in state canal construction waned in the 1850s, the Whigs retained a greater belief in the state's older canal policies and the Democrats split along radical and conservative lines.[75]

In Indiana, Governors James B. Ray (1825-31) and Noah Noble (1831-37) found that the canals virtually dominated their political life, even though Noble had supported the canal movement only cautiously and Ray had favored railroads over canals.[76] The legislature of 1835-36, which enacted the Mammoth Internal Improvement Act, was dominated by Whigs, but twenty-four out of thirty-seven Democrats voted for the bill. On the debate over the Butler bills in 1846-47, which were designed to satisfy the bondholders on the debt and enable the Wabash and Erie Canal to be completed, the Whigs supported them and the most concerted opposition came from within the Democratic party.[77] But Butler engineered his solution to the problem of Indiana's state debt with the support of the Democratic governor James Whitcomb.

In Illinois, the Whigs in the legislature of 1837 launched an ambitious program of canal and railroad construction that did not survive the depression of that year, and the continuation of the Illinois and Michigan Canal became the dominant issue in the election of 1842. The Whig candidate, Joseph Duncan, sought to explain an ambiguous statement on the issue, and the successful Democratic candidate, Thomas Ford, was accused of taking contradictory positions on the canal depending on whether he was speaking in the northern or southern parts of the state. Sectional interests were stronger than political philosophy, and the Whigs of southern Illinois generally opposed the canal.[78] Once in office, Governor Ford gave decisive assistance to the plan of Arthur Bronson in 1843 for resolving the debt crisis with eastern and foreign bondholders and completing the Illinois and Michigan Canal. This required passage of a tax bill, which was supported by Michael Ryan, the Democratic chairman of a committee on canals in the legislature. Although sectional interests determined most votes in the legislature, the opposition to Bronson's plan came primarily from Democrats in southern Illinois, and more Whigs than Democrats gave the votes that allowed the Illinois and Michigan Canal to go forward.[79]

The great canals built in Maryland and Virginia by private companies and aided by these states were equally influenced by partisan conflict. President Charles F. Mercer of the Chesapeake and Ohio Canal took the Jacksonian veto of the Maysville Road bill in 1830 as a constitutional attack on the measure of 1826 by which Congress had first assisted his company's canal. The "great national waterway" received no further congressional assistance, and it became increasingly subject to partisan conflict in Maryland.[80]

The election in 1838 of a Democratic governor, William Grayson, resulted in the removal of the Whig canal board on the Chesapeake and Ohio Canal and its replacement in 1839 by Democratic members, who faced the problem of liquidating the canal debt. But the new Democratic president of the board, Francis Thomas, differed little from his Whig predecessors such as Mercer in seeking to find ways to continue canal construction to the west. The chief effect of the party turnover, as in Ohio, was in the application of the spoils system under the Democratic board. A wholesale removal of older canal officers, including the clerk of the canal company, the treasurer, key engineers, and other officials, hindered canal construction. In 1841, however, another shift in the political balance in Maryland brought a new Whig board, which promptly rehired many of the former canal officials. The election of Governor Thomas E. Pratt was influenced by the lobbying efforts of J. H. B. Latrobe and George Peabody, who were financed in part by the Baring Company of London in its efforts to secure a resumption of interest payments in Maryland.[81] Still, the Chesapeake and Ohio Canal Company faced a constant shortage of funds in the 1840s and 1850s and finally ended construction above Harpers Ferry.

Partisan differences exercised a major influence in the state appropriations for the company that sought to complete the James River and Kanawha Canal in Virginia. Wayland Fuller Dunaway, the chief historian of the canal, called it the "football of politics."[82] The canal suffered because, though the Whigs in the Virginia legislature supported the canal, the Democrats were more frequently in control of the state government. As the canal moved west into the mountains, its greatest support was found in the western counties, where the Whigs were strongest. It came to be regarded as a "Whig enterprise," even though "the hearts of the people in Democratic strongholds were made glad by substantial appropriations to internal improvements in those sections."[83]

A Whig-controlled House in the legislature of 1834-35 made the appropriations that allowed the James River and Kanawha Canal Company to take over the canal in 1835, but subsequent Democratic legislatures provided only limited funds for construction. In 1842 a Democratic legislature enacted restrictive legislation that brought financial stringency to company operations and finally halted work on the canal. Efforts of a Whig House in the legislative session of 1844-45 to make a substantial appropriation to the canal company were blocked in the Democratic Senate. The legislature finally appropriated funds to allow the company to carry the canal beyond Lynchburg, but it required that construction stop at Buchanan. Just before the Civil War the Democratic

Governor Henry A. Wise attempted with little success to strengthen the canal project as an effort to bind western Virginia more closely to the tidewater.[84]

Localism and sectionalism in almost every canal state embroiled the canals in complex political rivalries, especially in the 1830s. The southern tier of counties in New York, little benefited by the Erie Canal, voted against the upper New York counties on every canal issue. A classic local conflict divided the interests of Peter B. Porter and his brother Augustus Porter at Black Rock on the Niagara River against the interests of Samuel Wilkeson at Buffalo on Lake Erie (villages only a few miles apart) over the issue of the western terminus for the Erie Canal. The lateral canals in New York were built in response to local political interests and logrolling, leading to a costly, overextended system that could not support itself in the economic crisis of the 1840s.

In Illinois, the same sectional divisions divided that state over the Illinois and Michigan Canal. Southern counties, little benefited by the canal, opposed it and northern counties united to gain the needed legislative support. In Ohio political trade-offs determined the two routes for trunkline canals, one to benefit the southwest corner of the state along the Miami River from Dayton to Cincinnati, the other to combine two more easterly lines between Cleveland and Akron and then between Columbus and Portsmouth. Localism lay behind the doctrine of "equal benefits" that brought an egalitarian expansion to the Ohio canals and the system of mixed enterprise in the Loan Law in 1837.

In Indiana the more thickly settled southeastern corner of the state won the little Whitewater Canal leading from Lawrenceburg on the Ohio River north to Brookville and Cambridge City, and that canal exercised a controlling influence on the fortunes of the far longer Wabash Canal project. Together they led to the Mammoth Internal Improvement program of 1836 and another vastly overextended system. The abortive Central Canal grew out of the rival local interests of Indianapolis, which was bypassed by the older Wabash trade route to the north.

Localism in Pennsylvania was based on the mountainous geography of the state, with the people of rival river valleys all seeking an outlet for their trade. In particular, this rivalry produced the Erie Extension Canal and the French Creek feeder in the northwestern corner of the state. But the greatest influence of localism was manifest in the Canal Act of 1827, which diverted state funds away from the Mainline Canal to more local projects. This localism in state after state was most prevalent in the canal construction of the 1830s, and it fueled the pejorative judgments conveyed by such terms as *canal craze* or *canal mania*.

Land grants were a direct stimulus to the expansion of the canal system. Long before that designation was used regarding railroads, state and federal aid produced land-grant canals. Four and a half million acres of land were donated in two series. In the 1820s and 1830s Ohio, Indiana, Illinois, and Wisconsin received land grants for canals, and these were followed by grants to Michigan and Wisconsin in the 1850s and 1860s. These grants were particularly significant for the completion of the Miami Extension Canal in northwestern Ohio and were decisive in leading Indiana to extend the Wabash and Erie Canal to Evansville. The grant to Illinois paid for five-sixths of the Illinois and Michigan Canal.[85]

The practice of giving alternate sections of land (a section was one square mile) on either side of the canal, designed both to aid the canals and to raise the value of the public lands, set the model for the railroad grants after the Civil War. The record of the sales of these lands was mixed. The Ohio Canal land grants were a clear benefit to northwestern Ohio when it was still a near wilderness. They included 438,000 acres given for the Miami Extension Canal from Dayton to the Wabash and Erie Canal, 292,000 acres for the Ohio portion of the Wabash and Erie Canal, and a 500,000-acre "floating" grant of vacant public land near the canals. Together they yielded more than $2 million, but this was only a small fraction of the cost of the Miami and Erie Canal. Moreover, the 1.25 million acres of land granted to Ohio were sold under conditions of fraud, mismanagement, and conflict between Ohio and the General Land Office. Ohio acted in such an insular fashion that, according to Scheiber, "there might as well have been a Great Wall of China" separating the disposition of the Ohio land grants from the experience of other states.[86]

Three grants to Indiana including 766,000 acres in the Vincennes Land District were well managed, and their sale yielded some $3 million. These grants influenced the ill-fated decision to extend the Wabash and Erie Canal to Evansville on the Ohio River and facilitated the second Butler Bill of 1847 for settling the state debt. Wisconsin received a quarter-million-acre grant but never completed the canal for which it was donated between the Fox and Wisconsin rivers. The grant for the Illinois and Michigan Canal was most successful, yielding $5.8 million, almost the entire cost of the canal. In Michigan, the major land grant of 75,000 acres was given by the state to a private company in a mutually beneficial arrangement; the company received the land for speculation, and the state acquired a completed canal at the Sault.[87]

Construction by states and assistance from federal land grants made American canals largely a product of public enterprise. Yet the emerging corporation and private entrepreneurial leadership made many canals the work of mixed enterprise. Ralph D. Gray's study of the Chesapeake and Delaware Canal describes a canal that was built in Maryland and Delaware by a company chartered in those states and in Pennsylvania as well, promoted through the efforts of Mathew Carey, and aided by two congressional stock subscriptions.[88] A similar admixture is found in the financial support for the Chesapeake and Ohio Canal. Pennsylvania's public works received broad popular support, which led to state subscriptions of stock in numerous canal and navigation companies. But the money was borrowed, and the stock subscriptions became part of the $40.9 million state debt by 1846. Accordingly, the act for the sale of the public works in 1857 included the prohibition of any state subscriptions to corporations in the form of mixed enterprise.[89]

The accomplishments of private companies in canal building may have been slighted in the attention accorded the great state enterprises that were celebrated as the fruits of republican government. The Louisville and Portland Canal Company in Kentucky accomplished the long-awaited task of bypassing the falls in the Ohio and made possible the upriver passage of the steamboat. The Sault Canal at the straits in Michigan was of equal significance for facilitating the trade of the upper Great Lakes. In the East various anthracite canals, the Lehigh canal in Pennsylvania, the Delaware and Hudson Canal in Pennsylvania and New York, and the Delaware and Raritan and Morris canals in New Jersey all were among the most successful canals of the Canal Era. All were built by private companies, though many enjoyed public assistance in the form of corporation privileges, lotteries, or banking privileges.

One of the smaller private company canals was the three-mile Milan Canal from the lake port of Huron to Milan in Ohio, which made Milan a great interior wheat port. The short canal routes from the Ohio and Erie Canal to the Ohio River were the work of private enterprise. The Pennsylvania and Ohio Canal, also called the Mahoning Canal, ran from Warren and Youngstown in Ohio to New Castle on the Erie Extension Canal; and to the south, the Sandy and Beaver Canal ran seventy-three miles from Bolivar on the Ohio and Erie Canal to East Liverpool on the

Ohio River. The short Cincinnati and Whitewater Canal, privately built, ran only a few miles past the home of Willian Henry Harrison at North Bend in Ohio to Indiana, but it diverted much of the trade on the Whitewater Valley to Cincinnati.

In the mixed enterprise that built and managed American canals there were men in both state and private projects who combined public interest with private gain. Fear of monopoly, embezzlement, or fraud was strong in Jacksonian values, but the modern concept of conflict of interest had not yet emerged. Often the individuals who identified the progress of these canals with their own fortunes seem almost larger than life. They expected personal gain, but their application of energy to public development went well beyond self-interest. They may have lacked advanced education but often knew law and surveying, they had financial experience, and they either possessed or rapidly acquired detailed knowledge of their states. Few were politicians. The progress of the canals for which they labored often hinged on their personal reputations, apart from any political office they may have held.

In the 1790s such a Federalist and Democratic-Republican elite had backed the pioneer canals, men such as General Philip Schuyler in New York, James Winthrop and Christopher Gore in Massachusetts, Francis Moultrie and John Rutledge in South Carolina, George Washington and Edmund Randolph in Virginia, and Robert Morris and Joshua Gilpin in Pennsylvania. They and others gave canals a share of their public service that has been better known in the political arenas of their states and the nation.

In the years of planning for the Erie Canal, Joseph Ellicott, land agent for the Holland Land Company, benefited the public with his knowledge of western New York and his warnings about the patterns of western trade which found its outlet through Montreal. But he also made sure that the canal followed a route in western New York that would give maximum benefit for the sale of the Holland Company's lands that he managed.[90] That there were limits to the blending of public service and private gain, however, was shown in the downfall of Myron Holley as a canal commissioner in the years of construction of the Erie Canal. He served faithfully as paymaster to contractors up and down the canal with minimal bureaucratic support, but when he embezzled funds for private land speculation along the canal, punishment was swift and severe, and he lost everything.

The driving force of a group of canal titans in the first half of the nineteenth century was in many ways comparable to that of the industrial entrepreneurs in the decades after the Civil War. As has been described in previous chapters, Josiah White drove forward the private project of the Lehigh Canal and the public construction of the Delaware Extension Canal in Pennsylvania, as did Mathew Carey on the Chesapeake and Delaware Canal, Arthur Bronson on the Illinois and Michigan Canal, John W. Brooks on the Sault Canal in Michigan, and Samuel Hanna and Charles Butler on the Indiana canals. Alfred Kelley and Micajah T. Williams were the persistent partners who projected and completed much of the Ohio canal system, and both built business empires founded on their canal labors. Kelley went into banking and railroads, moving his residence from Cleveland to the state capital of Columbus. Williams also moved into banking and founded the Ohio Life Insurance and Trust Company, which helped finance the Miami Extension Canal and the Wabash and Erie Canal. In 1842 Kelley staked his personal credit on the credit of the state and journeyed to England to sell bonds. If these men gave almost heroic service to bring their canal projects into existence, they did so in the face of the crippling stringencies of the Panic of 1837 during the very years when the canal network was becoming overextended.

Publicly and privately constructed canals reached 3,326 miles by 1840 and 4,254 miles by 1860, though by the latter year some 350 miles of canal had been abandoned.[91] In this era of agricultural expansion, canals gave the great staple crops of the interior an outlet to eastern ports, much as the steamboat did for the export of southern cotton. Merchandise went west for the growing population of the interior. Yet the overbuilding of canals in the 1830s began to be deprecated as a "canal mania" only a decade or less after the completion of the Erie Canal in 1825. The golden age when the canals appear as shining examples of public or private enterprise seems brief, and in the midst of the depression of 1837, many canals that were begun in a flush of optimism ended in financial failure.

Their financial failure was only one aspect, if a major one, of the collapse of the American economy after the panic that began in 1837. Wider problems, many of them originating outside the United States, weakened the Jacksonian economy in these years, and recovery did not come until late in the 1840s. By 1842 nine states had defaulted on their bonds, many for debts incurred in the building of canals. Two years stand in starkest economic contrast in the states, 1836 and 1842. In 1836 New

York began to enlarge the Erie Canal (authorized in 1835); Ohio moved toward its Loan Law, which assisted private canal companies; the Mammoth Internal Improvement Bill was passed in the Indiana legislature; and construction began on the Illinois and Michigan Canal. In 1842, however, stop laws brought canal construction to a halt, or nearly to a halt, in New York on the Erie Canal, in Maryland on the Chesapeake and Ohio Canal, in the canals of Indiana, Virginia, and most other canal-building states. Construction would resume, however, in the late 1840s and early 1850s.

Harvey H. Segal finds three cycles of canal investment. The first, from 1815 to 1834, peaked in 1828 and included the trunk lines in New York, Ohio, Pennsylvania, and Maryland. The second, from 1834 to 1844, included the rapid expansion of the lateral canals, stimulated by inflationary commodity prices and foreign loans and ending with the downturn after the depression of 1837. In this period the Pennsylvania and Ohio canal systems were joined; important canal links such as the Susquehanna and Tidewater Canal were added in the East; the Chesapeake and Ohio Canal was extended; the James River and Kanawha was being improved; and the enlargement of the Erie Canal was begun. Segal's third cycle continued from 1845 to 1860 and was devoted to completing unfinished canals in the face of rising competition from railroads. The success or failure of a particular canal was influenced by where it appeared in these cycles.[92]

Another economic analysis has distinguished between developmental canals, built ahead of settlement and creating their own commerce, and exploitative canals such as the New Jersey canals that connected active commercial centers.[93] Thus canals that may not have been profitable in yielding direct returns such as tolls earned for the state or a company were profitable in a long-run developmental sense, as has been claimed for the Indiana canals.

By almost any standard the Erie Canal was the great success of the Canal Era. Its tolls repaid its costs within five years; it helped to open western New York to settlement and aided New York City's growth; and it survived the drain of unprofitable lateral canals.[94] Similarly, the Ohio and Erie Canal and the Miami Canal from Dayton to Cincinnati have been judged successful, especially in the developmental sense.[95] So too have the anthracite canals that carried Pennsylvania coal to Philadelphia and New York City been judged a financial success, and they continued in service longer than any others. In 1868 Henry V. Poor, an early transportation historian, employed a broad definition of commercial success and selected the Erie Canal, the Chesapeake and Delaware Canal, and the Delaware and Raritan Canal as the only commercially successful canals in the United States.[96]

Julius Rubin judged the Pennsylvania Mainline and its larger lateral system an economic failure, an opinion shared by many historians. The Chesapeake and Ohio Canal could not compete with the Baltimore and Ohio Railroad, which was begun at almost the same time, and stopped construction at Harpers Ferry, far short of the Ohio. The Indiana canals were an unquestioned failure in direct financial returns, yet they may have been a developmental success. Even the Illinois and Michigan Canal on the vital connection between Chicago and the Mississippi was judged economically unprofitable in direct returns.[97]

However uneven their profitability, the canals were a dominant feature of the Farmer's Age in America, and their primary function was to carry agricultural staples to eastern markets. Where the ton-mile cost was between 15 and 25 cents by wagon, costs for carrying agricultural staples were reduced to 2 cents or less per ton-mile, varying by canal and product. For example, in 1850 on the Ohio canals, which have been examined most closely in this regard, wheat was carried for 1.4 cents and flour, lard, and pork for 1.2 cents. On the Erie Canal, the ton-mile rate for flour was 1.3 cents in 1851, and on the Pennsylvania Mainline Canal the rate for flour was 1.8 cents in 1849.[98] Significantly, the most rapid decline in costs came in the 1840s and 1850s, before the influence of the railroad could be felt extensively. State canal commissioners were under constant pressure to reduce tolls in these decades. Some economic historians have found this cost reduction the basis for regional specialization with the development of an agricultural West that supplied a manufacturing Northeast. But in a continuing debate, others have argued that the canals encouraged both agriculture and manufactures in the West, thus leading to diversification rather than specialization.[99]

If the primary function of the canals was to transport the products of the land, there is irony in their impact on urban centers, and their progress was inseparable from that of the cities they served. The rise of the port of New York was stimulated by products brought down the Hudson from the Erie Canal, by imported manufactures destined to travel west by canal, and by immigrants entering the port who would travel by canal boat to the Great Lakes and beyond. If too much is claimed for the influence of the Erie Canal, Robert Greenhalgh Albion has found that the "cotton triangle" and New York's "capitalization of its sea routes" in world trade were greater stimuli for the rise of New York port.[100]

An early urban historian, Blake McKelvey, described the Erie Canal as the "mother of cities," creating the new cities of Syracuse, Rochester, and Buffalo. More recently, Stuart Blumin has examined the city of Kingston on the lower Hudson as a port crossing the "urban threshold" after the completion of the Delaware and Hudson Canal. Hubertis Cummings has described a network of canal ports in Pennsylvania, which included Easton, Bristol, Columbia, Lewisburg, Muncy, Lewiston, Beaver, and New Castle. Other towns or cities that also served canal trade were Harrisburg, Liverpool, Berwick, Blairville, Franklin, and Meadville. Canals helped to make Reading Pennsylvania's third largest city by 1860. Akron and Toledo were among the new cities in Ohio created by the canals; Cincinnati grew to be the Queen City of the West stimulated by the Miami and Erie Canal; and Cleveland became the largest city in Ohio by 1850 as the terminus of the Ohio and Erie Canal. Before Chicago grew as a railroad center, its population multiplied through the influence of the Illinois and Michigan Canal; and in Indiana, a long line of canal ports from Fort Wayne to Evansville on the Ohio River was created by the Wabash and Erie Canal.[101]

Within these cities, the process of urban development was continually shaped by the passage of a canal. For example, Vivienne Dawn Maddox has described the development of four Erie Canal villages as they became cities: Syracuse, Palmyra, Rochester, and Lockport. In these cities canal basins were like the urban train stations of a later age; canal bridges determined the links between neighborhoods; vertical grain elevators were juxtaposed against the horizontal line of the canal; and Greek revival houses enhanced the classical tone so evident in the Romanesque arches of aqueducts as beautiful as the one that still stands at Palmyra.[102] Moreover, wherever canals were enlarged, rebuilt, or repaired, which was more frequent than has been indicated in the chapters above, canal construction caused the tearing up and rebuilding of streets, houses, and businesses, especially as the efforts to keep the older canal operating crossed the construction of the new.

Closer examination of the influence of canals on urban growth has gone beyond description of a line of canal ports to study the impact of the canal on the city and its hinterland, following in particular the urban patterns suggested by Eric Lampard. A study by Roberta B. Miller shows the growth of Syracuse at the junction of the Erie and Oswego canals in New York, but its agricultural hinterland declined in the face of competition from farther west. The recent study of Hamilton, Ohio, on the Miami and Erie Canal, gives a contrasting picture of hinterland growth related to the larger city of Cincinnati as the canal terminus. Perhaps most significant is the study of urban growth in Philadelphia by Diane Lindstrom, which shows that city growing hand in hand with its hinterland, aided by seven regional canals. The dramatic growth of Philadelphia also indicates a return from the Pennsylvania canal system far greater than the profit-and-loss record mentioned above might suggest.[103]

The losses in direct returns from the Pennsylvania canal system, which left the state with expenditures of $68 million against tolls of only $25 million, have led to the belief that Philadelphians and other Pennsylvanians erred grievously when they failed to turn to the new technology of the railroad. With the success of the Erie Canal an almost certain boon for the rival port of New York, Philadelphia, like Boston and Baltimore, had faced the great decision whether to imitate New York with rival canals or follow those who saw in the railroad the future of transportation to the West.[104] Indeed, Pennsylvania had sent the canal engineer William Strickland to England to examine both canals and railroads, and he returned to recommend the new technology of the railroad over canals. But Pennsylvania had chosen canals, and the costly Mainline system followed.

Boston, however, decided to delay, and when the fervor for canals had faltered at the prospect of a tunnel through the Berkshires, a railroad reached across the New England mountains to Albany in 1851. In New England the railroad was the agency of transportation development.[105] The New Haven and Northampton Canal might have been expected to have carried the trade of the Connecticut River Valley south from New Hampshire to the sea, but it failed miserably, and a railroad paralleled the Connecticut River by 1841.

Baltimore resisted the clamor for canals and audaciously inaugurated the Baltimore and Ohio Railroad even before the feasibility of such construction was assured. Baltimore thus outflanked both the Pennsylvania canals and the abortive western route of the Chesapeake and Ohio Canal, leaving the projected union of the latter waterways on the Ohio River at Pittsburgh only a forlorn dream. Yet the completion of the Baltimore and Ohio Railroad to the Ohio River was not as distinct from canals as it may appear. Between Point of Rocks and Harpers Ferry the railroad and the Chesapeake and Ohio Canal ran side by side through the narrow gorge of the Potomac, fighting each other in a prolonged litigation for right-of-way. Meanwhile, Maryland took virtual financial control of the Chesapeake and Ohio Canal after 1835, and Baltimore merchants successfully pursued the Susquehanna Valley trade by the Susquehanna and Tidewater Canal.

Yet the success of the Erie Canal demonstrated the potential of canals on a vast scale in America, after years of development as shorter waterways in England, France, or the Netherlands. At the peak of the first cycle of canal building in 1828, when the great western trunk lines were being constructed, railroads could not yet surmount steep grades, still used wooden or iron-strapped rails, and could not carry heavy freight. It was not until 1829 that Robert Stephenson's *Rocket* proved the locomotive power of the steam engine at the famous Rainhill Trials on the Liverpool and Manchester Railway in England. For a decade western states faced the same decision of whether to build canals or railroads that had faced eastern rivals to the Erie Canal. As late as 1836, Governor Noah Noble took the conservative choice in Indiana and completed the Wabash and Erie Canal. Considering the long political process involved in authorizing public canals and problems of capital, labor, and construction, it is little wonder that few were willing to gamble in the late 1820s on the new technology of the railroad.

When railroads were first introduced into the United States, their relationship to canals was symbiotic. In New York, the Mohawk and Hudson Railroad began in 1831 to carry passengers the seventeen miles from Albany to Schenectady to avoid the day-long passage through twenty-one locks, and it carried no freight. In 1837 the Camden and Amboy Railroad was combined with the Delaware and Raritan Canal in New Jersey. On the route of the Pennsylvania Mainline Canal, the Philadelphia and Columbia Railroad was opened in 1834 and ran eighty-four miles to link up with the Mainline Canal at Columbia on the Susquehanna River. When it began, it used both steam power and horsepower. At the crest of the Alleghenies on the Mainline route the thirty-seven-mile Alleheny Portage Railroad pulled section boats on cars over the mountains. In northeastern Pennsylvania, the Delaware and Hudson Canal was extended into the mountains for eighteen miles on the Delaware and Hudson Gravity Railroad. In 1829, the Stourbridge Lion was brought from England, through the Delaware and Hudson Canal, to be used briefly at Honesdale. Josiah White designed the Lehigh and Susquehanna Railroad to run between Wilkes-Barre and White Haven in Pennsylvania, and it extended twenty-five miles between his company's canal on the Lehigh River and the Susquehanna North Branch Canal and was partially opened in 1843.

It was in the second cycle of canal building, during the 1830s, that railroad construction spread rapidly in America. Three railroads out of Boston were chartered in 1830-31: one north to Lowell, one south to Providence, and one west to Worcester, and all were finished by 1835. In South Carolina the railroad from Charleston to Hamburg, across the Savannah River from Augusta, Georgia, had its first steam locomotive in 1830, and this 136-mile route became the longest in the nation when it was completed in 1833. The Baltimore and Ohio Railroad reached Harpers Ferry in 1836, well ahead of the Chesapeake and Ohio Canal.

The pattern of building railroads on the level stretches parallel to canals began early. Ten little railroad lines, including the Schenectady and Utica, the Utica and Syracuse, and the Rochester and Batavia, linked Albany and Buffalo before they were combined in 1853 as part of the New York Central Railroad. Many railroads were built as the plums of logrolling, ironically a charge first leveled at canals. Thus the Erie Railroad was finished in 1851 to satisfy the southern tier of counties in New York, which were not benefited by the Erie Canal. The great canal expansion in Ohio, Indiana, and Illinois in the 1830s included provision for compensatory railroads to satisfy rival local interests and resulted from the internal improvements measures enacted in these states in 1836 and 1837.

Canal engineers such as John B. Jervis of New York and Loammi Baldwin, Jr., of Massachusetts went on to become distinguished railroad engineers. Jervis started as an axman and surveyor on the Erie Canal near Rome, New York, and later designed the Croton Aqueduct to supply water to New York City. As a railroad engineer he designed the Mohawk and Hudson Railroad and invented the flexible truck for the forward wheels of a locomotive. Baldwin worked with his father on the Middlesex Canal, worked on the Pennsylvania canals, and then served as an engineer on railroads in Pennsylvania and Michigan. In Pennsylvania, Josiah White began as the moving force behind the company-built canals in the Lehigh Valley and then added railroad projects which included the ingenious gravity "Switchback Railroad" between Summit Hill and Mauch Chunk. John Edgar Thompson moved from work on the Pennsylvania canals to become a president of the Pennsylvania Railroad. Western canal builders such as Alfred Kelley, Micajah Williams, Samuel Hanna, and John W. Brooks made a similar transition to railroad interests.

Urban growth in canal cities changed with the transition to the dominance of the railroad. At the terminus of the Illinois and Michigan Canal, Chicago grew more rapidly when it became the great railroad hub of the Midwest. Cincinnati became the great city of the lower Ohio River as the terminus of the Miami and Erie Canal but lost out to Louisville when the Louisville and Nashville Railroad offered better connections to the South. Fort Wayne, however, made a steady transition from the major city on the Wabash and Erie Canal to the railroad center of northeastern Indiana.

By the 1850s the railroad came to dominate east-west trade, and economic development from transportation, as Albert Fishlow has shown, was largely the result of railroad expansion.[106] Passenger travel dropped on the canals in the 1840s, and freight followed in the 1850s. Railroad consolidations opened through routes from eastern cities to Chicago and St. Louis. The New York Central Railroad was consolidated in 1853, the Erie Railroad reached Dunkirk on Lake Erie in 1851, the Pennsylvania Railroad was completed to Pittsburgh in 1852, and in the latter year the Baltimore and Ohio Railroad reached Wheeling on the Ohio River. To protect their canals, New York and other states sought until 1851 to charge a toll on products carried on railroads when the canals were closed, and canal tolls were cut. But lack of revenue only speeded the decline of the canals.

Yet the supremacy of the railroad must be qualified. The anthracite canals continued to thrive, carrying bulky, heavy coal cheaper than could be done by rail. And more tonnage moved down the Erie Canal to the East in 1854 than on all of the eastern railroads combined. Its peak year would come in 1872.[107]

The canals reached their peak while they were being overtaken by the railroad, and in many ways the canals themselves had helped create the new technology that superseded them. Railroad engineers gained experience in dealing with levels and inclines needed as much for railroad roadbeds and grades as for waterways. The mixed enterprise of the Canal Era stimulated railroads even as it expanded canals. Government assistance to canals became the model for similar aid to railroads, offering public sanction for private development, setting precedents for land grants, and producing entrepreneurs well schooled in justifying private benefit for public good. Although there is little agreement on the contribution of canals to economic development, railroads entered an economy well advanced from the economy in which transportation needs had been served by rivers, roads, and wagons.

To attempt to grasp the full dimension of the Canal Era one must consider the thousands of miles of waterways that were constructed and operated over five decades by an innovative generation who made canals central to their lives. They used their governments and organized companies to build a canal network that often served them well until depression struck or the railroad passed by their canals. These canals depended upon the will of the electorate and the philosophy of those in power. Their expansion reflected a philosophy of egalitarianism, and whether they were public or private, canals were always dependent in some degree upon the democratic political process. Political leaders often tied their fortunes to canals; and the canals were the work of men who were almost heroic in the energy and skill they devoted to them. The result was the successful application of a new canal technology to America. And with a minimal bureaucracy, the canals were operated in a regional or national system that added a new element to the infrastructure of American transportation.

The Canal Era also created canals of drama and beauty, which can be only partially glimpsed today. Across them moved a mass of humanity, on line boats that carried both passengers and freight and on packets that ran night and day and in the 1840s carried a hundred people or more. They brought the bustle of the canal basin and new excitement to isolated rural settlements. To carry the millions of tons of agricultural staples listed cumulatively in statistical summaries, thousands of freighters each carried sixty to seventy-five tons, or as many as a thousand barrels of flour or four thousand bushels of wheat, and they made uncounted lockages as they changed levels on the canals. These long, narrow, silently moving boats passed between close banks and crossed over majestic aqueducts giving a sense of smooth, effortless speed. The long lines of canals now drawn on our maps and the canal structures still preserved are but the skeletal remains of a once vital dimension of American life.

Notes

1. Pioneer Canals and Republican Improvements

1. Charles Hadfield, *British Canals: An Illustrated History,* 7th ed. (London, 1984), chaps. 5-9.

2. Lewis Mumford, *Technics and Civilization* (New York, 1934), 15; John R. Stilgoe, *Common Landscape of America, 1580 to 1845* (New Haven, 1892), 115-21; Leo Marx, *The Machine in the Garden: Technology and the Pastoral Ideal in America* (New York, 1964), chap. 2; *Facts and Observations in Relation to the Origin and Completion of the Erie Canal* (New York, 1825), 3.

3. Darwin H. Stapleton, *The Transfer of Early Industrial Technologies to America* (Philadelphia, 1987), chap. 2; Richard Shelton Kirby, "William Weston and His Contribution to Early American Engineering," *Transactions of the Newcomen Society for the Study of the History of Engineering and Technology* 16 (1937): 1-17. Reprinted in Publications of the School of Engineering, Yale University, no. 26 (Sept. 1937).

4. Robert H. Wiebe, *The Opening of American Society: From the Adoption of the Constitution to the Eve of Disunion* (New York, 1984), 9.

5. General works on American canals include the following: Henry S. Tanner, *A Description of the Canals and Railroads of the United States* (New York, 1840); Henry Varnum Poor, *History of the Railroads and Canals of the United States of America,* 3 vols. (New York, 1860); Caroline E. MacGill et al., *History of Transportation in the United States before 1860* (Washington, D.C., 1917); Alvin Harlow, *Old Towpaths: The Story of the American Canal Era* (New York, 1926); Madeline Sadler Waggoner, *The Great Canal Era, 1817-1850* (New York, 1958); Harry Sinclair Drago, *Canal Days in America: The History and Romance of Old Towpaths and Waterways* (New York, 1972); George Rogers Taylor, *The Transportation Revolution, 1815-1860* (New York, 1951); and Carter Goodrich, *Government Promotion of American Canals and Railroads* (New York, 1960).

6. Stapleton, *Transfer of Industrial Technologies,* 40-41; see also J. Lee Hartman, "Pennsylvania's Grand Plan of Post-Revolutionary Internal Improvement," *Pennsylvania Magazine of History and Biography* 65 (October 1941): 445-54.

7. Quoted in Hartman, "Philadelphia's Grand Plan," 446.

8. Thomas C. Cochran, ed., *The New American State Papers, Transportation,* 7 vols. (Wilmington, Del., 1972), 1:185-94 (hereafter cited as *NASP*); J.G. Francis, "The Union Canal," *Lebanon County Historical Society Publications* 2 (1939): 245-57.

9. John F. Bell, "Robert Fulton and the Pennsylvania Canals," *Pennsylvania History* 9 (July 1942): 191-96. For a picture of an inclined plane drawn by Fulton and accounts of early American engineers, see William H. Shank and others, *Towpaths to Tugboats: A History of American Canal Engineering* (York, Pa., 1982), 11-22.

10. *NASP, Transportation,* 141-45.

11. James W. Livingood, "The Canalization of the Lower Susquehanna," *Pennsylvania History* 8 (April 1941): 132-34; see also W. Curtis Montz, "Water Transportation on the Susquehanna," *Wyoming Historical and Geological Society Publications* 22 (1970): 21-25, and Gerald Smeltzer, *Canals along the Lower Susquehanna, 1796-1900* (York, Pa., 1963), 7-9.

12. James Weston Livingood, *The Philadelphia-Baltimore Trade Rivalry, 1780-1860* (Harrisburg, 1947), 34.

13. *NASP, Transportation,* 1:263; Stapleton, *Transfer of Industrial Technologies,* 59-62.

14. Washington to the governor of Virginia, Oct. 10, 1784, in John C. Fitzpatrick, ed., *The Writings of George Washington,* 39 vols. (Washington, D.C., 1931-44), 27:471-81.

15. Jefferson to Washington, March 15, 1784, in Julian P. Boyd, ed., *The Papers of Thomas Jefferson,* 20 vols. (Princeton, 1950-82), 7:26-27.

16. Douglas R. Littlefield, "The Potomac Company: A Misadventure in Financing an Early American Internal Improvement Project," *Business History Review* 58 (Winter 1984): 583; see also Douglas R. Littlefield, "Maryland Sectionalism and the Development of the Potomac Route to the West, 1786-1826," *Maryland Historian* 14 (Fall–Winter, 1983): 31-52.

17. See Alex Crosby Brown, *The Dismal Swamp Canal* (Hilton Village, Va., 1945).

18. Dixon Ryan Fox, *The Decline of Aristocracy in the Politics of New York* (New York, 1919), 51.

19. Nathan Miller, "Private Enterprise in Inland Navigation: The Mohawk Route prior to the Erie Canal," *New York History* 31 (Oct. 1950): 398-413.

20. Ronald E. Shaw, *Erie Water West: A History of the Erie Canal, 1792-1854* (Lexington, 1966), 14-21.

21. Weston's MS Notebook, Institution of Civil Engineers, London.

22. David Maldwyn Ellis, *Landlords and Farmers in the Hudson-Mohawk Region, 1790-1850* (Ithaca, 1946), 84; Shaw, *Erie Water West,* 15-21.

23. Quoted in Shaw, *Erie Water West,* 20.

24. Christopher Roberts, *The Middlesex Canal, 1793-1860* (Cambridge, Mass., 1938), 42-44, 182; Thomas C. Proctor, "The Middlesex Canal: Prototype for American Canal Building," *Canal History and Technology Proceedings* 7 (March 1988): 125-72. (Hereafter cited as CHTP.)

25. Daniel Hovey Calhoun, *The American Civil Engineer: Origins and Conflict* (Cambridge, Mass., 1960), 17; quotation from Proctor, "Middlesex Canal," 139-40.

26. Quoted in Proctor, "Middlesex Canal," 145. Calhoun, *American Civil Engineer,* 94-99, 103; see Frederick A. Abbott, "The Role of the Civil Engineer in Internal Improvements: The Contribution of the Two Loammi Baldwins, Father and Son, 1776-1838" (Ph.D. dissertation, Columbia University, 1952).

27. Baldwin to Weston, [March] 1794, quoted in Elting E. Morison, *From Know-How to Nowhere: The Development of American Technology* (New York, 1974), 24.

28. Roberts, *Middlesex Canal,* 99.

29. Ibid., 62.

30. Ibid., 160-69.

31. Loammi Baldwin, Jr., to James F. Baldwin, Dec. 14, 1807, quoted in Abbott, "Civil Engineer," 55.

32. Roberts, *Middlesex Canal,* 164-70.

33. Quotation from Dirk J. Struik, *Yankee Science in the Making* (Boston, 1948) 114. Edward C. Kirkland, *Men, Cities, and Transportation: A Study in New England History, 1820-1900,* 2 vols. (Cambridge, Mass., 1948) 1:65.

34. Henry D. Thoreau, *A Week on the Concord and Merrimack Rivers,* ed. Carl F. Houde et al. (Princeton, 1983), 62, 258.

35. Roberts, *Middlesex Canal,* 183.

36. Henry Savage, Jr., *River of the Carolinas: The Santee* (Chapel Hill, 1956), 111-23.

37. *NASP, Transportation,* 1:103-5; Savage, *River of the Carolinas,* 242.

38. Gates to Jefferson, Sept. 24, 1780, in Boyd, ed., *Papers of Thomas Jefferson,* 3:662; Jefferson to Speaker of the House of Delegates, May 10, 1781, ibid., 5:627-28.

39. Ulrich Bonnell Phillips, *A History of Transportation in the Eastern Cotton Belt to 1860* (New York, 1908), 40-41.

40. Quoted in Savage, *River of the Carolinas,* 242.

41. *NASP, Transportation,* 1:104-5.

42. Phillips, *Transportation in the Eastern Cotton Belt,* 41-43.

43. Robert Mills, *Atlas of the State of South Carolina* (Columbia, 1825); Mills, *A Treatise on Inland Navigation* (Baltimore, 1820); Mills, *Internal Improvement of South Carolina: Particularly Adapted to the Low Country* (Columbia, 1822); *NASP, Transportation,* 1:105-12; MacGill et al., *Transportation in the United States,* 257.

44. F.A. Porcher, *The History of the Santee Canal* (Charleston, 1903), n.p.

45. Helen Mar Pierce Gallagher, *Robert Mills, Architect of the Washington Monument, 1781-1855* (New York, 1935), 134-55; Mills, *Atlas*; Mills, *Inland Navigation.*

46. For examples, see Miller, "Private Enterprise in Inland Navigation," 402-13; and Littlefield, "Potomac Company," 581.

47. John F. Kasson, *Civilizing the Machine: Technology and Republican Values in America, 1776-1900* (New York, 1976), 50.

48. Quoted in Proctor, "Middlesex Canal," 149.

49. Quoted in J.G. Francis, "The Union Canal," *Lebanon County Historical Society Publications* 2 (1939): 240-43.

50. Hartman, "Philadelphia's Grand Plan," 453.

51. Harold C. Syrett, ed., *The Papers of Alexander Hamilton,* 27 vols. (New York, 1961-89), 5:310; John R. Nelson, Jr., *Liberty and Property: Political Economy and Policymaking in the New Nation, 1789-1812* (Baltimore, 1987), 46.

52. Jacob Cooke, *Tench Coxe and the Early Republic* (Chapel Hill, 1978), 81, 402-3, 489-90; quotation on 489. See especially Tench Coxe, *An Enquiry into the Principles on Which a Commercial System for the United States Should Be Provided* (Philadelphia, 1787).

53. Coxe to Jefferson, Nov. 8, 1801, quoted in Cooke, *Tench Coxe,* 403.

54. Andrew R. L. Cayton, *The Frontier Republic: Ideology and Politics in the Ohio Country, 1780-1825* (Kent, Ohio, 1986), 146-52.

55. Drew R. McCoy, *The Elusive Republic: Political Economy in Jeffersonian America* (New York, 1980), 163.

56. Nelson, *Liberty and Property,* 124-25.

57. Boyd, ed., *Papers of Thomas Jefferson,* 3:662, 668, 11:446, 454.

58. John Lauritz Larson, "'Bind the Republic Together': The National Union and the Struggle for a System of Internal Improvements," *Journal of American History* 74 (Sept. 1987): 371; see also Joseph H. Harrison, Jr., "*Sic et Non*: Thomas Jefferson and Internal Improvement," *Journal of the Early Republic* 7 (Winter 1987): 339.

59. Jefferson, Second Inaugural Address, in James D. Richardson, ed., *A Compilation of the Messages and Papers of the Presidents, 1789-1847* 10 vols. (Washington, D.C., 1896-1903), 1:379.

60. Joseph H. Harrison, Jr., describes this as "Jeffersonian nationalism." See Harrison, "*Sic et Non,*" 341-42; see also Harrison, "The Internal Improvement Issue in the Politics of the Union, 1783-1825" (Ph.D. dissertation, University of Virginia, 1954); Robert Fulton, *A Treatise on the Improvement of Canal Navigation* (London, 1796); Joel Barlow, *The Vision of Columbus,* in William K. Bottorff and Arthur I. Ford, eds., *The Works of Joel Barlow,* 2 vols. (Gainesville, 1970), 1:346.

61. Lee W. Formwalt, "Benjamin Henry Latrobe and the Revival of the Gallatin Plan of 1808," *Pennsylvania History* 48 (April 1981): 106-12; Ralph D. Gray, *The National Waterway: A History of the Chesapeake and Delaware Canal, 1769-1965* (Urbana, Ill., 1967), 25-27. A new edition of Gray, *National Waterway,* was published in 1989, covering the expansion and operation of the canal, 1965-1985.

62. Gray, *National Waterway,* 25-28; Larson, "'Bind the Republic Together,'" 372; Harrison, "Internal Improvement Issue," 188-99.

63. Gallatin's *Report on Roads and Canals* is in *NASP, Transportation,* 1:17-275; Carter Goodrich, "The Gallatin Plan after One Hundred and Fifty Years," *American Philosophical Society Proceedings* 102 (Oct. 1958): 437; Larson, "'Bind the Republic Together,'" 372-74; Nelson, *Liberty and Property,* 125-26.

64. Larson, "'Bind the Republic Together,'" 374.

65. Formwalt, "Latrobe and the Gallatin Plan," 114.

66. Harrison, "Internal Improvement Issue," 300.

67. Madison, Seventh Annual Message, in Richardson, ed., *Messages and Papers,* 1:567-68.

68. Roger H. Brown, *The Republic in Peril: 1812* (New York, 1964), 67-87; Wiebe, *Opening of American Society,* 200.

69. Madison, Eighth Annual Message, in Richardson, ed., *Messages and Papers,* 2:576.

70. *Annals of Congress,* 14th Cong., 2d sess., 851-960.

71. Ibid., 15th Cong., 1st sess., 1371-72.

72. Madison to Monroe, Nov. 29, 1817, in Gaillard Hunt, ed., *The Writings of James Madison,* 9 vols. (New York, 1900-1910), 8:397.

73. *Annals of Congress,* 14th Cong., 2d sess., 106-61.

74. Larson, "'Bind the Republic Together,'" 382-83. Drew McCoy has written that Madison believed he had long made his constitutional reservations clear and that he had "adverted specifically to the need for a constitutional amendment." In the Bonus Bill Madison "perceived nothing less than a threat to constitutional government itself" (McCoy, *The Last of the Fathers: James Madison and the Republican Legacy* [Cambridge, 1989], 94, 97).

75. *Annals of Congress,* 14th Cong., 2d sess., 880.

76. Ibid., 865.

77. Ibid., 15th Cong., 1st sess., 1175.

78. Ibid., 1198-99.

79. Ibid., 1389.

80. De Witt Clinton to Rufus King, December 13, 1817 Rufus King Papers, New-York Historical Society; Shaw, *Erie Water West,* 67-78.

81. Quoted in Gray, *National Waterway,* 29.

82. Wiebe, *Opening of American Society,* chap. 10.

2. Great Lakes to Atlantic

1. Calhoun, *American Civil Engineer,* 34-37, 105, 122.

2. David Maldwyn Ellis, "The Rise of the Empire State, 1790-1820," *New York History* 56 (Jan. 1975): 6.

3. Shaw, *Erie Water West,* 24.

4. Hawley essays, *Genesee Messenger,* 1807-8, quoted in Shaw, *Erie Water West,* chap. 2. For the originating role of Gouverneur Morris, see Julius Rubin, "An Innovating Public Improvement: The Erie Canal," in Carter Goodrich, ed., *Canals and American Ecomonic Development* (New York, 1961), 26-28.

5. Hawley essays, *Genesee Messenger,* 1807-8, quoted in Shaw, *Erie Water West,* 27-28.

6. Ibid., 28.

7. *Laws of the State of New York, in Relation to the Erie and Champlain Canals,* 2 vols. (Albany, 1825), 1:44.

8. Stapleton, *Transfer of Industrial Technologies,* 69-71.

9. Shaw, *Erie Water West,* 53-55.

10. Ibid., chap. 19; Roger Evan Carp, "The Erie Canal and the Liberal Challenge to Classical Republicanism, 1785-1850" (Ph.D. dissertation, University of North Carolina at Chapel Hill, 1986), 53-55.

11. Clinton, "Memorial of the citizens of New-York, in favor of a Canal Navigation . . . February 21, 1816," *Laws of the State of New York,* 1:122-41.

12. Shaw, *Erie Water West,* chap. 4.

13. Carp, "Erie Canal and Republicanism," 233-47.

14. *Rochester Telegraph,* April 25, 1820.

15. Shaw, *Erie Water West,* 118-19, chap. 9; see Alvin Kass, *Politics in New York State, 1800-1830* (Syracuse, 1965).

16. New-York Corresponding Association for the Promotion of Internal Improvements, *Public Documents, Relating to the New-York Canals* (New York, 1821), xlii; see also Calhoun, *American Civil Engineer,* 24-37.

17. Crozet's career at West Point is described in John C. Greene, *American Science in the Age of Jefferson* (Ames, Iowa, 1984), 131-34. See also Richard Shelton Kirby et al., *Engineering in History* (New York, 1956), 207-20.

18. Shaw, *Erie Water West,* 87-91.

19. Ibid., 93, 119-20. Convicts from the Auburn State Prison were employed on the western section of the canal in 1821 (ibid., 90-91, 98, 129). Carp emphasizes ethnic tensions from increasing employment of Irish laborers and convicts ("Erie Canal and Republicanism," 344-46).

20. *Laws of the State of New York,* 1:450, 2:63.

21. Leland R. Johnson, *The Davis Island Lock and Dam, 1870-1922* (Pittsburgh, 1985), 56.

22. Contemporary response to the appearance of the Erie Canal is examined in detail in Carp, "Erie Canal and Republicanism," 704-34.

23. Clinton to David Thomas, June 10, 1827, David Thomas Papers, New York State Library, Albany.

24. Henry Seymour to De Witt Clinton, Oct. 16, 1821, De Witt Clinton Papers, Columbia University Library, New York.

25. Shaw, *Erie Water West,* 280; David Moldwyn Ellis, "Albany and Troy—Commercial Rivals," *New York History* 24 (Oct. 1943): 484-511.

26. Shaw, *Erie Water West,* 192.

27. *Niles' Weekly Register* 29 (Oct. 1, 1825): 66.

28. Page Smith, *The Shaping of America: A People's History of the Young Republic,* 4 vols. (New York, 1980), 3:773-74.

29. Cadwallader D. Colden, *Memoir . . . at the Celebration of the Completion of the New York Canals* (New York, 1825), Appendix.

30. Colden, *Memoir,* 5.

31. Paul Wallace Gates, *The Farmer's Age: Agriculture, 1815-1860* (New York, 1960), 254; Shaw, *Erie Water West,* 264-65.

32. Figures on the grain shipped from Buffalo and arriving at the Hudson on the Erie Canal are found in Shaw, *Erie Water West,* chaps. 14-15; figures for grain arriving at Buffalo or Oswego from the Great Lakes are found in John G. Clark, *The Grain Trade in the Old Northwest* (Urbana, Ill., 1966), chap. 5.

33. Shaw, *Erie Water West,* 298.

34. Ibid., 294-95.

35. Construction on the original Erie Canal was completed in 1825 with dimensions of forty feet by four feet; enlargement to dimensions of seventy feet by seven feet was authorized in 1835 but not completed until 1862; the present Erie Barge Canal, which canalized the rivers through which it passed, was begun in 1905. For the fullest account of the engineering aspects and trade of the Erie Canal see Noble E. Whitford, *History of the Canal System of the State of New York Together with Brief Histories of the Canals of the United States and Canada,* 2 vols. (Albany, 1906).

36. Ibid., 1:909-10.

37. Ronald W. Filante, "A Note on the Economic Viability of the Erie Canal, 1825-1860," *Business History Review* 48 (Spring 1974): 96-97. See also J.R. Ringwalt, *Development of Transportation Systems* (New York, 1888), 47.

38. Filante, "Note," 98-99.

39. Thomas McIlwraith, "Freight Capacity and Utilization of the Erie and Great Lakes Canals before 1850," *Journal of Economic History* 36 (Dec. 1976): 865-67. See also Hugh G.J. Aitken, *The Welland Canal Company* (Cambridge, Mass., 1954).

40. Shaw, *Erie Water West,* 298.

41. Quoted from the *Genesee Farmer* in the *Rochester Daily Advertiser,* June 9, 1832.

42. *Rochester Telegraph,* June 28, 1825.

43. John Disturnell, *A Guide between Washington, Baltimore, Philadelphia, New York and Boston* (New York, 1846), 59.

44. Quoted in Shaw, *Erie Water West,* 293.

45. Thomas X. Grasso, "Rochester's Canal Boat Industry," *Bottoming Out,* May 1987, pp. 9-18.

46. Filante, "Economic Viability of the Erie Canal," 101.

47. Whitford, *History of the Canal System of New York,* 2:1064-68.

48. Kirkland, *Men, Cities and Transportation,* 1:60.

49. Ibid., 81-84; see Blackstone Canal Company Business Records, 1796-1899, folder 2, 1826-50, American Antiquarian Society Library, Worcester, Mass.

50. William Green Roelker, "The Providence Plantations Canal," *Rhode Island History* 5 (Jan. 1946): 23.

51. James B. Hedges, *The Browns of Providence Plantations,* 2 vols. (Providence, 1968), 2:210-16.

52. Vincent Edward Powers, "'Invisible Immigrants': The Pre-famine Irish Com-

munity in Worcester, Massachusetts, from 1826 to 1860" (Ph.D. dissertation, Clark University, 1976), 99-113.

53. Ibid., 105-8, 112-20.

54. Hedges, *The Browns of Providence Plantations,* 2:213-14. See also Alden Gould, "History of the Blackstone Canal, 1828-1848," *American Canals* no. 9, (May 1974): 3-4.

55. Kirkland, *Men, Cities and Transportation,* 1:83-84.

56. Hedges, *The Browns of Providence Plantations,* 2:216.

57. Kirkland, *Men, Cities and Transportation,* 1:72-73; Abbott, "Civil Engineer," 137.

58. Kirkland, *Men, Cities and Transportation,* 1:73-75.

59. Taylor, *Transportation Revolution,* 38.

60. Charles Rufus Harte, *Connecticut's Canals* (Hartford, 1938), 3-4; James Mark Composeo, "The History of the Canal System between New Haven and Northampton, 1822-1847," *Historical Journal of Western Massachusetts* 6 (Fall 1977): 37-39. See also *An Account of the Farmington Canal Company; and of the Hampshire and Hampden Canal Company; and of the New Haven and Northampton Company till the Suspension of Its Canals in 1847* (New Haven, 1850).

61. Harte, *Connecticut's Canals,* 28.

62. Charles Rufus Harte, *Some Engineering Features of the Old Northampton Canal* (New Haven, 1933), 13-21. There were thirty-two locks and five aqueducts in Massachusetts and twenty-eight locks and three aqueducts in Connecticut.

63. Harte, *Old Northampton Canal,* 4; Composeo, "Canal System between New Haven and Northampton," 56-67.

64. Composeo, "Canal System between New Haven and Northampton." 44.

65. Harte, *Connecticut's Canals,* 35. See also William P. Donovan, "The New Haven and Northampton Canal," in Lawrence E. Wikander et al., eds., *The Northampton Book: Chapters from 300 Years in the Life of a New England Town, 1654-1954* (Northampton, 1954), 85-89.

66. Robert A. Ludwig, "An Economic Analysis of the New Haven and Northampton Canal," unpublished paper, Forbes Library, Northampton, Massachusetts, 27, 37-38; Harte, *Connecticut's Canals,* 26.

67. Harte, *Old Northampton Canal,* 3. See also J.M. Franceschi, "Hampshire and Hampden Canal," *American Canals* no. 35, (Nov. 1980): 7.

68. Abbott, "Civil Engineer," 136-42; Kirkland, *Men, Cities and Transportation,* 1:80.

69. Harlow, *Old Towpaths,* 70-73; Julius Rubin, "Canal or Railroad? Imitation and Innovation in the Response to the Erie Canal in Philadelphia, Baltimore, and Boston," *American Philosophical Society Transactions* 51, pt. 7 (Philadelphia, 1961), 80-86.

70. Kirkland, *Men, Cities and Transportation,* 1:84-90.

3. Mid-Atlantic Network

1. Louis C. Hunter, *Steamboats on the Western Rivers* (Cambridge, Mass., 1949); Erik F. Haites, James Mak, and Gary M. Walton, *Western River Transportation: The Era of Internal Development, 1810-1860* (Baltimore, 1975), 91-94; Edith McCall, *Conquering the Rivers: Henry Miller Shreve and the Navigation of America's Inland Waterways* (Baton Rouge, 1984), 132-34, 166-72; Paul B. Trescott, "The Louisville and

Portland Canal Company, 1825-1874," *Mississippi Valley Historical Review* 44 (March 1958): 686-708.

2. Livingood, *Philadelphia-Baltimore Trade Rivalry,* 1-2; Hartman, "Pennsylvania's Grand Plan," 441, 447; Richard Nelson Pawling, "Geographic Influences upon the Development and Decline of the Union Canal," *Proceedings of the Canal History and Technology Symposium* 2 (March 1983): 70, 72 (hereafter cited as *PCHTS*).

3. Pawling, "Union Canal," 372.

4. Abbott, "Civil Engineer," 95.

5. The engineers were Simeon Guilford, George T. Olmstead, and Sylvester Welch. See Calhoun, *American Civil Engineer,* 97-99.

6. Pawling, "Union Canal," 73.

7. Ibid., 74-80; Livingood, *Philadelphia-Baltimore Trade Rivalry,* 108-11. The Union Canal was enlarged in 1857.

8. Livingood, "Canalization of the Lower Susquehanna," 145-49; Montz, "Water Transportation on the Susquehanna," 29; Smeltzer, *Canals along the Lower Susquehanna,* 97-99.

9. Rubin, "Canal or Railroad?" 3.

10. Julius Rubin, "An Imitative Public Improvement: The Pennsylvania Mainline," in Goodrich, ed., *Canals and Development,* 69.

11. W. Bernard Carlson, "The Pennsylvania Society for the Promotion of Internal Improvements: A Case Study in the Political Uses of Technological Knowledge, 1824-1826," *CHTP* 8 (March 1988): 179-92.

12. Ralph D. Gray, "Philadelphia and the Chesapeake and Delaware Canal, 1769-1823," *Pennsylvania Magazine of History and Biography* 84 (Oct. 1960): 408-22.

13. Louis Hartz, *Economic Policy and Democratic Thought: Pennsylvania, 1776-1860* (Cambridge, Mass., 1948), chap. 4.

14. Richard C. Shelling, "Philadelphia and the Agitation in 1825 for the Pennsylvania Canal," *Pennsylvania Magazine of History and Biography* 62 (April, 1938): 182, 197.

15. Quoted in Rubin, "Imitative Public Improvement," 85.

16. Avard Longley Bishop, "The State Works of Pennsylvania," *Publications of Yale University* (New Haven, 1907), 185, (reprinted from the *Transactions of the Connecticut Academy of Arts and Sciences* 13 [November 1907]).

17. Ibid., 192; Robert McCullough and Walter Leuba, *The Pennsylvania Mainline Canal* (York, Pa., 1973), 51.

18. John C. Trautwine, Jr., "The Philadelphia and Columbia Railroad of 1834," *Philadelphia History* 2 (1925): 160.

19. Albright Zimmerman, "The Columbia and Philadelphia Railroad: A Railroad with an Identity Problem," *CHTP* 3 (1984): 72-74. The railroad used two inclined planes, one 2,805 feet long near the eastern end and one 1,800 feet long at the western end. The latter was abandoned in 1840 (Bishop, "State Works of Pennsylvania," 196).

20. Hubertis M. Cummings, "James D. Harris, Principal Engineer, and James S. Stevenson, Canal Commissioner," *Pennsylvania History* 18 (Oct. 1951): 293-301.

21. Solomon W. Roberts, "Reminiscences of the First Railroad over the Allegheny Mountain," *Pennsylvania Magazine of History and Biography* 2 (1878): 371-72.

22. Robert D. Ilsevich and Carl K. Burkett, Jr., "The Canal through Pittsburgh: Its Development and Physical Character," *Western Pennsylvania Historical Magazine* 68 (Oct. 1985): 360-66.

23. "Grant's Hill Canal Tunnel," *Canal Currents* no. 40 (Fall 1977): 6-8.

24. Nicklin, quoted in McCullough and Leuba, *Pennsylvania Mainline Canal,* 44. The locks here were ninety by fifteen feet, two feet narrower than on the Eastern Division.

25. Welch had designed the locks on the Eastern Division. For the career of Moncure Robinson, see Stapleton, *Transfer of Industrial Technologies,* 128-40.

26. Hubertis M. Cummings, "John August Roebling and the Public Works of Pennsylvania," ed. Donald Sayenga, *CHTP* 4 (1984): 101-7; Donald Sayenga, *Ellet and Roebling* (York, Pa., 1983), 20-21; John P. Miller and others, "Roebling," *Canal Currents* no. 61 (Winter 1983): 1-16; Roberts, "Reminiscences," 378-81. The inclined planes were designed by W. Milnor Roberts.

27. Roberts, "Reminiscences," 374-75.

28. Ibid., 377, 380.

29. Charles Dickens, *American Notes for General Circulation,* 2 vols. (1842; rpt. New York, 1972), 199.

30. Nicklin, quoted in William H. Shank, *The Amazing Pennsylvania Canals* (York, Pa., 1960), 8.

31. Jesse L. Hartman, "John Dougherty and the Rise of the Section Boat System," *Pennsylvania Magazine of History and Biography* 69 (Oct. 1945): 294-314. Canvass White recommended such boats for the portage railroad in 1826 (C. P. Yoder, "History's Greatest Canal Engineer," *Canal Currents* no. 10 [Fall 1969]: 4).

32. Hartman, "John Dougherty," 313. The container concept for transferring cargoes was attempted, but the experiment failed. "Safety cars" were invented to prevent accidents on the inclines, and there were few runaway cars.

33. Rubin, "Imitative Public Improvement," 107. Total cost through 1857 was $16,472,633 (Poor, *Railroads and Canals,* 1:558).

34. Rubin, "Imitative Public Improvement," 108.

35. George B. Johnson, "Whiskey Shipped on the Western Division," *Canal Currents* no. 67 (Summer 1984): 7.

36. Harlow, *Old Towpaths,* 127-28.

37. Earl J. Heydinger, "Packets, 'Huntington to Philadelphia,'" *Canal Currents,* no. 65 (Winter 1984): 7.

38. Hubertis M. Cummings, "Pennsylvania: Network of Canal Ports," *Pennsylvania History* 21 (July 1954): 270.

39. Rubin, "Imitative Public Improvement," 111.

40. Poor, *Railroads and Canals,* 1:558; Bishop, "Pennsylvania State Works," 229; Willard R. Rhodes, "The Pennsylvania Canal," *Western Pennsylvania Historical Magazine* 4 (Sept. 1960): 217.

41. Hartz, *Economic Policy and Democratic Thought,* 143.

42. Albright G. Zimmerman, "Governments and Transportation Systems: Pennsylvania as a Case Study," *CHTP* 6 (March 1987): 33, 35.

43. Ibid., 27.

44. Rubin, "Imitative Public Improvement," 107; Bishop, "Pennsylvania State Works," 228.

45. The full extent of the Pennsylvania canals is described in T.B. Klein, *The Canals of Pennsylvania and the System of Internal Improvements* (Harrisburg, 1901), and Bishop, "Pennsylvania State Works." For the canals in the northwestern area of the state, see Lloyd A.M. Corkan, "The Beaver and Lake Erie Canal," *Western Pennsylvania Historical Magazine* 17 (Sept. 1934): 175-88.

46. Detailed maps on a scale of sixteen miles to the inch are found in Christopher T. Baer, *Canals and Railroads of the Mid-Atlantic States, 1800-1860* (Wilmington, Del., 1981); Hartz, *Economic Policy and Democratic Thought,* 151.

47. Bishop, "Pennsylvania State Works," 229.

48. Thomas C. Cochran, *Pennsylvania: A Bicentennial History* (New York, 1978), 119.

49. Basic sources for the anthracite canals are Chester L. Jones, *The Economic History of the Anthracite-Tidewater Canals,* University of Pennsylvania Publications in Political Economy and Public Law (Philadelphia, 1908), abridged in "The Anthracite-Tidewater Canals," *Annals of the American Academy of Political and Social Science* 30 (1908): 102-16. See also T.K. Woods, "Anthracite and Slackwater," *PCHTS* 2 (March 1983): 45-67.

50. H. Benjamin Powell, "Coal and Pennsylvania's Transportation Policy, 1825-1828," *Pennsylvania History* 38 (April 1971): 134-51.

51. Walter S. Sanderlin, "The Expanding Horizons of the Schuylkill Navigation Company, 1815-1870," *Pennsylvania History* 36 (April 1969): 174-91.

52. Powell, "Coal and Pennsylvania's Transportation Policy," 144-45.

53. F. Charles Petrillo, *Anthracite and Slackwater: The North Branch Canal, 1828-1901* (Easton, Pa., 1986), chap. 3; Woods, "Anthracite and Slackwater," 46-48; Leroy Bugbee, "The North Branch Canal," Wyoming Historical and Geological Society, *Proceedings and Collections* 24 (1984), 68-69.

54. Petrille, *Anthracite and Slackwater,* 90-91.

55. Bugbee, "North Branch Canal," 91.

56. Ibid., 76; see also Woods, "Anthracite and Slackwater," 48-51. Edward J. Davies II, however, found the North Branch Canal less important in the Wyoming Valley in the early years. He noted that coal output in 1841 and 1842 in the Wyoming Valley was only 4 to 5 percent of the production of coal in eastern Pennsylvania, and the valley lagged behind neighboring coal regions (*The Anthracite Aristocracy: Leadership and Social Change in the Hard Coal Regions of Northeastern Pennsylvania, 1800-1930* [DeKalb, Ill., 1985], 18).

57. Hubertis M. Cummings, "James D. Harris and William B. Foster, Jr., Canal Engineers," *Pennsylvania History* 24 (July 1957): 197.

58. Ibid., 191.

59. Sidney Davis. "The West Branch Canal," *Proceedings of the Northumberland Historical Society,* 26 (1967), 28-43.

60. Montz, "Water Transportation on the Susquehanna," 30; Cummings, "Pennsylvania," 266.

61. Montz, "Water Transportation on the Susquehanna," 29; George B. Scriven, "The Susquehanna and Tidewater Canal," *Maryland Historian* 71 (Winter, [1976]): 522-76; Livingood, "Canalization of the Lower Susquehanna," 145-49.

62. Edward Steers, "The Delaware and Hudson Canal Company's Gravity Railroad," *PCHTS* 2 (March 1983): 130-203; Jones, *Economic History of the Anthracite-Tidewater Canals,* chap. 4.

63. Neal FitzSimmons, ed., *The Reminiscences of John B. Jervis, Engineer of the Old Croton* (Syracuse, 1971), chap. 4, p. 402, n. 6.

64. Robert M. Vogel, *Roebling's Delaware and Hudson Canal Aqueduct* (Washington, D.C., 1971), 13; Sayenga, *Ellet and Roebling,* 33-34; Peter Osborne III, "The Delaware and Hudson Canal Company's Enlargement and the Roebling Connection,"

CHTP 2 (March 1984): 128. In 1848 John A. Roebling moved his wire mill from Pennsylvania to Trenton, New Jersey, close to the Delaware and Raritan Canal.

65. Jones, "Anthracite-Tidewater Canals," 103; Poor, *Railroads and Canals,* 1:355-56.

66. Sanderlin, "Schuylkill Navigation Company," 177; Anne Bartholomew, comp., and Lance E. Metz, researcher, *Delaware and Lehigh Canals* (Easton, Pa., 1989), 4-5.

67. Sanderlin, "Schuylkill Navigation Company," 185; Edward J. Gibbons, "The Building of the Schuylkill Navigation System, 1815-1828," *Pennsylvania History* 57 (Jan. 1990): 30-32; Jones, *Economic History of the Anthracite-Tidewater Canals,* chap. 7; Spiro G. Patton, "Charles Ellet, Jr., and the Canal vs. Railroad Controversy," *PCHTS* 2 (March 1983): 3-7.

68. Donald Sayenga, "The Untryed Business: An Appreciation of White and Hazard," *PCHTS* 2 (March 1983): 108-13; Thomas Dinkelacker, "The Construction of the Lehigh Canal and the Early Development of the Lehigh Valley Region," *Proceedings and Collections of the Wyoming Historical and Geological Society* 24 (1984): 45-64.

69. Sayenga, "Untryed Business," 108-25; E.J. Hartman, "Josiah White and the Lehigh Canal," *Pennsylvania History* 7 (Oct., 1940): 232-35.

70. Jones, "Anthracite-Tidewater Canals," 109; Jones, *Economic History of the Anthracite-Tidewater Canals,* chap. 2.

71. Alfred D. Chandler, Jr., "Anthracite Coal and the Beginnings of the Industrial Revolution in the United States," *Business History Review* 46 (Summer 1972): 152, 165, 179; Charles Waltman, "Influence of the Lehigh Canal on the Industrial and Urban Development of the Lehigh Valley," *PCHTS* 2 (March 1983): 87-104.

72. Taylor, *Transportation Revolution,* 40.

73. The Delaware Division Canal was later deepened to six feet. See C.P. Yoder, *Delaware Canal Journal: A Definitive History of the Canal and the River Valley through Which It Flows* (Bethlehem, Pa., 1972), 47.

74. Jones, "Anthracite-Tidewater Canals," 108; Jones, *Economic History of the Anthracite-Tidewater Canals,* chap. 3.

75. Yoder, *Delaware Canal Journal,* 221.

76. Horace Jerome Cranmer, "Internal Improvements in New Jersey: Planning the Morris Canal, 1822-1824," *Proceedings of the New Jersey Historical Society* 59 (Oct. 1951): 324-41; Calhoun, *American Civil Engineer,* 99-101.

77. Barbara N. Kalata, *A Hundred Years, A Hundred Miles: New Jersey's Morris Canal* (Morristown, N.J., 1983), 26-58, 107-11, 131-42, 219-31, 398-99; Wheaton J. Lane, "The Morris Canal," *Proceedings of the New Jersey Historical Society* 55 (Oct. 1937): 222-25. See James Lee, *The Morris Canal: A Photographic History* (Easton, Pa., 1979).

78. Kalata, *A Hundred Years, A Hundred Miles,* 641.

79. Lane, "Morris Canal," 226.

80. Ibid., 230, 252; Kalata, *A Hundred Years, A Hundred Miles,* 343-47, 370-73, 439.

81. H. Jerome Cranmer, "Improvements without Public Funds: The New Jersey Canals," in Goodrich, ed., *Canals and Development,* 142, 145.

82. Cranmer, "Internal Improvements in New Jersey," 333-41; Kalata, *A Hundred Years, A Hundred Miles,* 18-29, 79-83, 121-23, 388-93, 418-21; Michael Birkner, *Samuel Southard: Jeffersonian Whig* (Rutherford, N.J., 1984), 170-74.

83. Cranmer, "Improvements without Public Funds," 16-26, 147-61.

84. Ibid., 141-43, 148.

85. Robert T. Thompson, "Transportation Combines and Pressure Politics in New Jersey, 1833-1836," *Proceedings of the New Jersey Historical Society* 57 (Jan. 1939): 1. The company gave the state of New Jersey $200,000 worth of stock and paid $30,000 in annual transit fees (Poor, *Railroads and Canals,* 1:386).

86. Thompson, "Transportation Combines and Pressure Politics in New Jersey," 84-86.

87. Elizabeth G. C. Menzies, *Passage between Rivers: A Portfolio of Photographs with a History of the Delaware and Raritan Canal* (New Brunswick, N.J., 1976), 40; James and Margaret Cawley, *Along The Delaware and Raritan Canal* (Rutherford, N.J., 1970), 21.

88. J. Roscoe Howell, "Ashbel Welch, Civil Engineer," *Proceedings of the New Jersey Historical Society* 79 (Oct. 1961): part 1, 251-63; 70 (Jan. 1962): part 2, 46-53. When the canal was enlarged it was deepened to eight feet.

89. Cranmer, "Improvements without Public Funds," 156. Among the investors were John Potter of Charleston, South Carolina, John Jacob Astor, James Neilson, Robert F. Stockton and John C. Stevens.

90. Harlow, *Old Towpaths,* 209-10; Poor, *Railroads and Canals,* 1:386-89; MacGill, *History of Transportation,* 230-34.

91. Cranmer, "Improvements without Public Funds," 157-60.

4. The Chesapeake and Southern Canals

1. Walter S. Sanderlin, *The Great National Project: A History of the Chesapeake and Ohio Canal,* Johns Hopkins Studies in Historical and Political Science, vol. 64 (Baltimore, 1946). An earlier monograph is George Washington Ward, *The Early Development of the Chesapeake and Ohio Canal,* Johns Hopkins Studies in Historical and Political Science, vol. 17 (Baltimore, 1899).

2. Sanderlin, *Great National Project,* 57.

3. Allan Nevins, ed., *The Diary of John Quincy Adams* (New York, 1951), 382.

4. William M. Franklin, "The Tidewater End of the Chesapeake and Ohio Canal," *Maryland Historical Magazine,* (Winter 1986): 291-95; see Douglas Egerton, *Charles Fenton Mercer and the Trial of National Conservatism,* (Jackson, Miss. 1989) chap. 13.

5. John Lauritz Larson, "A Bridge, a Dam, a River: Liberty and Innovation in the Early Republic," *Journal of the Early Republic* 7 (Winter 1987): 368. Georgetown interests had seen the same advantage when they won a causeway-dam across the Potomac at Mason's Island opposite Georgetown, at the expense of Washington or Alexandria.

6. Franklin, "Chesapeake and Ohio Canal," 300.

7. Sanderlin, *Great National Project,* 84; Rubin, "Canal or Railroad?" 76-77.

8. Sanderlin, *Great National Project,* 106-12; Egerton, *Charles Fenton Mercer,* 385-86.

9. Sayenga, *Ellet and Roebling,* 26-37.

10. Ward, *Chesapeake and Ohio Canal,* 91-92. Montgomery C. Meigs and William R. Hutton were noted for their design and construction of the Cabin John Aqueduct near the point where John Quincy Adams had begun the canal. The aqueduct

had a 220-foot stone arch, the longest single-span stone arch in the Western Hemisphere at the time (Thomas F. Hahn, *The C & O Canal: An Illustrated History* [Sheperdstown, W.Va., 1980], 20; Hahn, *The Chesapeake and Ohio Canal: Pathway to the Nation's Capital* [Metuchen, N.J., 1984], chap. 6). Elizabeth Kytle described the work of Henry Richards, the company's representative in England, who recruited Welsh, English, and Irish workers (*Home on the Canal* [Cabin John, Md., 1983], 32). Robert Leckle came from South Carolina to become superintendent of masonry on the Chesapeake and Ohio Canal. A full account of Irish labor conflicts on this canal is found in Peter Way, "Shovel and Shamrock: Irish Workers and Labor Violence in the Digging of the Chesapeake and Ohio Canal," *Labor History* 30 (Fall 1989), 489-517.

11. Sanderlin, *Great National Project,* 113.

12. Richard B. Morris, "Andrew Jackson, Strikebreaker," *American Historical Review* 55 (Oct. 1949): 54-68; Way, "Shovel and Shamrock," 508-15.

13. Sanderlin, *Great National Project,* 127; Thomas F. Hahn, *Towpath Guide to the Chesapeake & Ohio Canal,* 4 vols., rev. ed. (Shepherdstown, W.Va., 1978), 4:40-45; Kytle, *Home on the Canal,* 57-58.

14. Copied in *National Intelligencer,* quoted in Hahn, *Towpath Guide,* 4:65-66.

15. Kytle, *Home on the Canal,* 66.

16. Ella E. Clark, ed., "Life on the C. & O. Canal: 1859," *Maryland Historical Magazine* 55 (March 1960): 110.

17. Sanderlin, *Great National Project,* 285.

18. Ibid., 205; Poor, *Railroads and Canals,* 1:604.

19. E.T. Coke, *A Subaltern's Furlough: Descriptive of the United States, Upper and Lower Canada, New Brunswick, and Nova Scotia, during the Summer and Autumn of 1832,* 2 vols. (New York, 1833), 1:92; Clark, "Life on the C. & O. Canal," 93.

20. Kytle, *Home on the Canal,* 102.

21. Ward, *Chesapeake and Ohio Canal,* 113.

22. Livingood, *Philadelphia-Baltimore Trade Rivalry,* 84-86; Gray, *National Waterway,* 1-15, 22.

23. Livingood, *Philadelphia-Baltimore Trade Rivalry,* 91.

24. Gray, *National Waterway,* 14, 55, 63-64.

25. Ibid., vii-xi. See also Ralph D. Gray, "Philadelphia and the Chesapeake and Delaware Canal, 1769-1823," *Pennsylvania Magazine of History and Biography* 84 (Oct. 1960): 401-23.

26. Calhoun, *American Civil Engineer,* 112.

27. Gray, *National Waterway,* 59-61.

28. Ibid., 66.

29. Tables are given in Livingood, *Philadelphia-Baltimore Trade Rivalry,* 98-99; see also Gray, *National Waterway,* 104-5.

30. Gray, *National Waterway,* 130; Livingood, *Philadelphia-Baltimore Trade Rivalry,* 98.

31. Gray, *National Waterway,* 119.

32. Livingood, "Canalization of the Lower Susquehanna," 138-47; Smeltzer, *Canals along the Lower Susquehanna,* 12-13, 42-61.

33. Poor, *Railroads and Canals,* 1:552-53; Livingood, *Philadelphia-Baltimore Trade Rivalry,* 72; Smeltzer, *Canals along the Lower Susquehanna,* 42.

34. Livingood, *Philadelphia-Baltimore Trade Rivalry,* 95-97, 110-14.

35. Ibid., 78-80.

36. Ibid., 94-97.

37. Wayland Fuller Dunaway, *History of the James River and Kanawha Company,* Columbia University Studies in History, Economics and Public Law, 104 (New York, 1922), 60, 63-68.

38. Calhoun, *American Civil Engineer,* 112; Charles Ellet, Jr., and Simon W. Wright, the son of Benjamin Wright, served as assistant engineers. The training of Moncure Robinson is described in Stapleton, *Transfer of Industrial Technologies,* 128-31. See also Dunaway, *James River and Kanawha Company,* 64-66.

39. Quoted in Abbott, "Civil Engineer," 151.

40. Dunaway, *James River and Kanawha Company,* 85, 88, 93, 119n.

41. T. Gibson Hobbes, Jr., "The James River and Kanawha Canal," in *The Best from American Canals,* no. 2, ed. Thomas F. Hahn and William E. Trout III (York, Pa., 1984), 40.

42. The company ultimately made a link to the Ohio River with 60 miles of channel improvements on the Kanawha River and 208 miles of turnpike roads over the mountains.

43. Dunaway, *James River and Kanawha Company,* 158, 165.

44. Ibid., 166.

45. Ibid.; Hobbes, "James River and Kanawha Canal," 422.

46. Dunaway, *James River and Kanawha Company,* 171-72; T. Gibson Hobbs, Jr., "Early Canal Boats on the James River and Kanawha Canal," *American Canals* no. 24 (Feb. 1978): part 1, 3, 6; no. 26 (Aug. 1978): part 2, 7.

47. Ibid., 183.

48. D.T. Bisbie, *An Appeal for the Continuation of the Water Line through Virginia* (Richmond, 1857).

49. Dunaway, *James River and Kanawha Company,* 198-203.

50. Carl Cahill, "Oldest Canal in Use in Peril" *Waterways Journal* 99 (May 20, 1985): 9; Tanner, *Description of the Canals and Railroads of the United States*, 161-62.

51. Clifford Reginald Hinshaw, Jr., "North Carolina Canals before 1860," *North Carolina Historical Review* 25 (Jan. 1948): 44. See also Alexander Crosby Brown, *The Dismal Swamp Canal* (Chesapeake, Va., 1967).

52. Alexander Crosby Brown, *Juniper Waterway: A History of the Albemarle and Chesapeake Canal* (Newport News, Va., 1981), 38.

53. Edmund Ruffin, *Agricultural, Geological and Descriptive Sketches of North Carolina* (1861), quoted in ibid., 41-42.

54. Albemarle and Chesapeake Canal Company, Fourth Annual Report (1859), quoted in ibid., 44.

55. Hinshaw, "North Carolina Canals," 55-56; Brown, *Juniper Waterway,* 119.

56. Ruffin, *Sketches of North Carolina,* quoted in Brown, *Juniper Waterway,* 41.

57. Hinshaw, "North Carolina Canals," 39.

58. Charles Clinton Weaver, *Internal Improvements in North Carolina Previous to 1860,* Johns Hopkins University Studies in Historical and Political Science, vol. 21, nos. 3-4 (Baltimore, 1903), part 1, chap. 5.

59. Porcher, *Santee Canal,* n.p.; Lewis W. Richardson, "The Canals of South Carolina," *American Canals* no. 4 (Feb. 1973): part 1, 6; no. 5 (May 1973) part 2, 4-5.

60. Milton Sydney Heath, *Constructive Liberalism: The Role of the State in Economic Development in Georgia to 1860* (Cambridge, Mass., 1954), 241. See also L.W. Richardson, "The Canals of Georgia," *American Canals* no. 6 (Aug. 1973): 6; no. 7 (Nov. 1973): part 2, 5; no. 8 (Feb. 1974): part 3, 4.

61. Phillips, *Transportation in the Eastern Cotton Belt,* 110-13.

62. Heath, *Constructive Liberalism,* 245.

63. T.G. Hobbes, Jr., "Edward Hall Gill, Civil Engineer," *Canal Currents* no. 49 (Winter 1980): 10-15.

64. Quoted in Abbott, "Civil Engineer," 213.

65. Thomas A. Becnel, *The Barrow Family and the Barataria and Lafourche Canal: The Transportation Revolution in Louisiana, 1829-1925* (Baton Rouge, La., 1989), chap. 2.

5. Canals of the Old Northwest

1. See the sources cited in Chapter 3 above. The present chapter draws from Ronald E. Shaw, "The Canal Era in the Old Northwest," in *Transportation and the Early Nation* (Indianapolis, 1982), 89-112.

2. Frank Wilcox, *The Ohio Canals,* ed. William A. McGill (Kent, Ohio, 1969), 1-8.

3. Harry N. Scheiber, *Ohio Canal Era: A Case Study of Government and the Economy, 1820-1861* (Athens, Ohio, 1969), 16-30; Scheiber, "The Ohio Canal Movement, 1820-1825," *Ohio Historical Quarterly* 69 (1960): 231-56; James L. Bates, *Alfred Kelley: His Life and Work* (Columbus, 1888), chaps. 2-3. The basic earlier study of the Ohio canals is C.P. McClelland and C.C. Huntington, *History of the Ohio Canals: Their Construction, Use and Partial Abandonment* (Columbus, 1905).

4. Scheiber, *Ohio Canal Era,* 357-58. See also Scheiber, "Alfred Kelley and the Ohio Business Elite, 1822-1859," *Ohio History* 87 (Autumn 1978): 365-92; "Entrepreneurship and Western Development: The Case of Micajah T. Williams," *Business History Review* 37 (Winter 1963): 345-68; and "Public Canal Finance and State Banking in Ohio, 1827-1837," *Indiana Magazine of History* 65 (June 1969): 129.

5. Scheiber, *Ohio Canal Era,* 53.

6. Ibid., 11, 130-33.

7. David A. Newhardt, "Miami and Erie Canal: Watering the Summit," *Towpaths* 23, no. 2 (1985): 13-24.

8. Scheiber, *Ohio Canal Era,* 126-28.

9. Frank W. Treverrow, "A Link to the Ohio River," *Towpaths* 24, no. 4 (1986): 21-24.

10. Harry N. Scheiber, "The Pennsylvania and Ohio Canal: Transport Innovation, Mixed Enterprise, and Urban Rivalry, 1825-1861," *Old Northwest* 6 (Summer 1980): 105-35. For the Erie Extension Canal, see Corkan, "The Beaver and Lake Erie Canal," 178-88.

11. Clark, *Grain Trade in the Old Northwest,* 67.

12. Scheiber, *Ohio Canal Era,* 193-95, 198-200, 238.

13. Clark, *Grain Trade in the Old Northwest,* 63.

14. Scheiber, *Ohio Canal Era,* 112, 222-23; Richard T. Farrell, "Internal-Improvement Projects in Southwestern Ohio, 1815-1834," *Ohio History* 80 (Winter 1971): 4-23; Terry K. Woods, "Early Trade on Ohio's Western Canal, 1827-1840," in *The Best from American Canals,* ed. Thomas F. Hahn, William H. Shank, and William E. Trout III (York, Pa., 1980), 45-46.

15. Daniel Preston, "Market and Mill Town: Hamilton, Ohio, 1795-1860" (Ph.D. dissertation, University of Maryland, College Park, 1987).

16. Ibid., 148-50.

17. Scheiber, "Pennsylvania and Ohio Canal," 105-35.

18. Scheiber, *Ohio Canal Era,* 112.

19. Clinton to Stickney, 1818, quoted in Elbert Jay Benton, *The Wabash Trade Route in the Development of the Old Northwest,* Johns Hopkins University Studies in Historical and Political Science, vol. 21, nos. 1-2 (Baltimore, 1903), 94. For an early study of the Indiana canals see Logan Esarey, *Internal Improvements in Early Indiana* (Indianapolis, 1912).

20. Paul H. Wehr, "Samuel Hanna: Fur Trader to Railroad Magnate" (Ph.D. dissertation, Ball State University, 1968).

21. Ralph D. Gray, "The Canal Era in Indiana," *Transportation and the Early Nation,* 116; Paul Fatout, *Indiana Canals* (West Lafayette, Ind., 1972), 39.

22. Fatout, *Indiana Canals,* 33, 40.

23. Gray, "Canal Era in Indiana," 118.

24. Fatout, *Indiana Canals,* 72; Benton, *Wabash Trade Route,* 43.

25. Fatout, *Indiana Canals,* 63; James M. Miller, "The Whitewater Canal," *Indiana Magazine of History* 3 (Sept. 1907): 108-15; Harry L. Rinker, "Whitewater Canal," *American Canals* no. 3 (Nov. 1972), n.p.

26. Fatout, *Indiana Canals,* 81-82.

27. Dennis K. McDaniel, "Water over Water: Hoosier Canal Culverts, 1832-1847," *Indiana Magazine of History* 78 (Dec. 1982): 310-11; "Aboite Creek Aqueduct: Above-Ground Archeology at Work," *Indiana Waterways,* Thomas Meek and Clarence Hudson, 5 (Winter 1988): 3-6.

28. Fatout, *Indiana Canals,* 93.

29. Taylor, *Transportation Revolution,* 47.

30. Gray, "Canal Era in Indiana," 121.

31. John Denis Haeger, *The Investment Frontier: New York Businessmen and the Economic Development of the Old Northwest* (Albany, 1981), 216-21.

32. James E. Fickle, "The 'People' versus 'Progress' in the Old Northwest: Local Opposition to the Construction of the Wabash and Erie Canal," *Old Northwest* 8 (Winter 1982-83): 309-28; Daniel W. Snepp, "Evansville's Channels of Trade and the Secession Movement, 1850-1865," *Indiana Historical Society Publications* 8 (1928): 340-58.

33. Taylor, *Transportation Revolution,* 47–48; Fatout, *Indiana Canals,* 48; Benton, *Wabash Trade Route,* 88.

34. Snepp, "Evansville's Channels of Trade," 384-87.

35. Clark, *Grain Trade in the Old Northwest,* 72.

36. Charles R. Poinsatte, *Fort Wayne during the Canal Era, 1828-1855,* Indiana Historical Society Collections, 46 (Indianapolis, 1969), 102.

37. Benton, *Wabash Trade Route,* chap. 3; Clark, *Grain Trade in the Old Northwest,* 72.

38. Gray, "Canal Era in Indiana," 129.

39. R. Carlyle Buley, *The Old Northwest: Pioneer Period, 1815-1840,* 2 vols. (Indianapolis, 1950), 2:261.

40. Miller, "Whitewater Canal," 108-15; Clark, *Grain Trade in the Old Northwest,* 17; *Hagerstown Exponent,* March 25, April 1, 15, 1981.

41. James William Putnam, *The Illinois and Michigan Canal: A Study in Economic History,* Chicago Historical Society Collections, 10 (Chicago, 1918), 29.

42. Catherine T. Tobin, "The Lowly Muscular Digger: Irish Canal Workers in Nineteeth Century America" (Ph.D. dissertation, University of Notre Dame, 1987), 81. See also Catherine T. Tobin, "Irish Labor on American Canals," CHTP 9 (March 1990): 2-35."

43. Board of Canal Commissioners, quoted in ibid., 47, 103, 138. John H.

Krenkel, *Illinois Internal Improvements, 1818-1848* (Cedar Rapids, Iowa, 1958), 46, 110, 125.

44. Haeger, *Investment Frontier,* 206-7. Krenkel, *Illinois Internal Improvements,* 181, 188-9.

45. Quoted in Tobin, "Lowly Muscular Digger," 47-48.

46. Ibid., 84-93.

47. Krenkel, *Illinois Internal Improvements,* 195.

48. Putman, *Illinois and Michigan Canal,* 102.

49. Samuel Mermin, *The Fox-Wisconsin Rivers Improvement: An Historical Study in Legal Institutions and Political Economy* (Madison, Wisc., 1968); Frederica Kleist, "Portage Canal History," and "Boats on the Portage Canal," in *Best from American Canals* 2 (1984) 66-67.

50. John N. Dickinson, *To Build a Canal: Sault Ste. Marie, 1853-1854 and After* (Columbus, 1981), 26-30, chaps. 5-10; Irene D. Neu, "The Building of the Sault Canal, 1825-1855," *Mississippi Valley Historical Review,* 40 (June 1953): 28ff.; Neu, "The Mineral Lands of the St. Mary's Falls Ship Canal Company," in David M. Ellis, ed., *The Frontier in American Development: Essays in Honor of Paul Wallace Gates* (Ithaca, 1969), 162-73.

51. Dickinson, *To Build a Canal,* 130.

52. F. W. Treverrow, "The Laphams—A Canal Family," *Towpaths* 27, no. 1 (1989): 1-5.

53. Dickinson, *To Build a Canal,* 80.

54. Quoted in ibid., 81.

55. Poinsatte, *Fort Wayne,* 65.

56. Stan Schmitt, "Census Records as a Canal Information Source," *Indiana Waterways* 3 (Winter 1984-85): 3-4.

57. David Burr to Noah Noble, Dec. 30, 1835, quoted in "'Canal Wars' 1835," *Indiana Waterways* 2 (Oct. 1982): 6-8.

58. Tobin, "Lowly Muscular Digger," 99-100.

59. Norma Brase, "Portrait of a Canal Boat Captain," *Indiana Waterways* 3 (Oct. 1983): 4.

60. Stan Schmitt, "An Evansville Canal Family," *Indiana Waterways* 4 (Spring 1985): 1-2.

61. Shaw, *Erie Water West,* 90; Ernest M. Teagarden, "Builders of the Ohio Canal, 1825-1832," *Inland Seas* 19 (1963): 95. As convicts were used on the Erie Canal, inmates of the Ohio State Penitentiary worked on the Columbus feeder of the Ohio and Erie Canal.

62. Quoted in Poinsatte, *Fort Wayne,* 222.

63. Ibid., chap. 5.

64. Scheiber, *Ohio Canal Era,* 237.

65. Terry K. Woods, "Crossing Ohio by Packet Boat: Passenger Packets on the Ohio & Erie Canal," *Gamut* no. 25 (Winter 1988): 76-82.

66. Ibid., 75.

67. Maurice Thompson, *Stories of Indiana* (New York, 1898), 219, letter of July 1851.

68. J. Richard Beste, *The Wabash: or Adventures of an English Gentleman's Family in The Interior of America,* 2 vols. (London, 1855), 2:193-219.

69. *Weekly Indiana State Sentinel,* Oct. 6, 1853, reprinted in *Indiana Waterways* 2 (April 1983): 4.

70. Poinsatte, *Fort Wayne,* 223-24; Scheiber, *Ohio Canal Era,* 237; John M. Lamb, "Early Days on the Illinois & Michigan Canal," *Chicago History* n.s. 3 (1974-75): 17.

71. Poinsatte, *Fort Wayne,* 230; Benton, *Wabash Trade Route,* 47.

72. Frederick Jackson Turner, *Rise of the New West, 1819-1829* (New York, 1906); Turner, *The Frontier in American History* (New York, 1920).

73. *Cincinnati Chronicle,* Dec. 8, 1827, quoted in *Towpaths* 25, no. 4 (1977): 40.

74. Donald T. Zimmer, "The Ohio River: Pathway to Settlement," *Transportation and the Early Nation,* 70-71.

75. Ibid., 73-74.

76. Eric F. Haites, James Mak, and Gary M. Walton, *Western River Transportation: The Era of Early Internal Development, 1810-1860,* Johns Hopkins University Studies in Historical and Political Science, vol. 93 (Baltimore, 1975), 21-22.

77. Clark, *Grain Trade in the Old Northwest,* 279.

78. Ibid., 88-90; Krenkel, *Illinois Internal Improvements,* 194-95.

79. Clark, *Grain Trade in the Old Northwest,* 105-18.

80. William F. Gephart emphasized the canals of the Middle West as "distinctively local transportation routes," which did not injure the New Orleans trade (*Transportation and Industrial Development in the Middle West,* Columbia University Studies in History, Economics and Public Law, 34 [New York, 1909], 118, 127). For more recent studies see Scheiber, *Ohio Canal Era,* 213-14; Clark, *Grain Trade in the Old Northwest,* chap. 7.

81. John D. Barnhart, *Valley of Democracy: The Frontier versus the Plantation in the Ohio Valley, 1775-1818* (Bloomington, 1953).

82. James Willard Hurst, *Law and the Conditions of Freedom in the Nineteenth Century United States* (Madison, Wisc., 1961). See Harry N. Scheiber, "At the Borderland of Law and Economic History: The Contributions of Willard Hurst," *American Historical Review* 75 (Feb. 1970): 744-56. For the Loan Law of 1837 see Scheiber, *Ohio Canal Era,* 130-33; for Indiana see Fatout, *Indiana Canals,* 72-73.

6. The Canal Network

1. Louis C. Hunter, *Water Power in the Century of the Steam Engine,* Vol. 1 of *A History of Industrial Power in the United States, 1780-1930* (Charlottesville, Va., 1979), 2.

2. Mumford, *Technics and Civilizations,* 120-23. Stilgoe, *Common Landscape of America,* 128. See also Brooke Hindle, ed., Material Culture of the Wooden Age (Tarrytown, N.Y. 1981), 9, 186.

3. Calhoun, *American Civil Engineer,* 22. In 1818 ten engineers made up the Topographical Bureau in the Engineer Department of the army under Secretary of War John C. Calhoun, and they engaged in civil as well as military operations. About two hundred civil engineers had graduated from West Point by 1837 (Forest G. Hill, *Roads, Rails and Waterways: The Army Engineers and Early Transportation* [Norman, Okla., 1957]). West Point engineers worked on the Chesapeake and Ohio Canal, the Morris Canal, the Delaware and Raritan Canal, and the Indiana canals (Stilgoe, *Common Landscape of America,* 124). For French influence on American engineering see Todd Shallat, "Building Waterways, 1802-1861: Science and the United States Army in Early Public Works," *Technology and Culture,* 31 (January 1990): 22-31, 36, 42, 46.

4. Morison, *From Know-How to Nowhere,* 23-25.

5. Stapleton, *Transfer of Industrial Technologies,* 71.

6. Calhoun, *American Civil Engineer,* 122; FitzSimmons, ed., *Reminiscences of John B. Jervis,* chaps. 2-7.

7. Calhoun, *American Civil Engineer,* chap. 5.

8. Ibid., 48-50; Raymond H. Merritt, *Engineering in American Society, 1850-1875* (Lexington, Ky., 1969), 40-42. The work of Erie Canal engineers on the New England canals is described in Camposeo, "Canal System between New Haven and Northampton," 37-53. Samuel Forrer has described his progression in Ohio from "junior rodman" to "Resident Engineer" on the Miami Canal, trained by Erie Canal engineers. His account is similar to that of John B. Jervis. See Samuel Forrer, "Notes for an Autobiography," n.d., Forrer-Pierce-Wood Collection, Dayton Public Library, Dayton, Ohio. For a list of engineers on American canals and a description of canal engineering see Shank, *Towpaths to Tugboats,* 31-39.

9. Scheiber, *Ohio Canal Era,* 69-71, 77-78, 125.

10. Powers, "'Invisible Immigrants,'" 105.

11. Tobin, "Lowly Muscular Digger," 101.

12. Way, "Shovel and Shamrock," 493.

13. Shaw, *Erie Water West,* 92; Scheiber, *Ohio Canal Era,* 77-78.

14. Scheiber, *Ohio Canal Era,* 71.

15. Myron B. Sharp, "Troubles on the Pennsylvania Canal," *Western Pennsylvania Historical Magazine* 52 (April 1969): 157.

16. Shaw, *Erie Water West,* 169-72.

17. *Report of the Select Committee of the Assembly of 1846, upon the Investigation of Frauds in the Expenditure of the Public Moneys upon the Canals of the State of New York* (Albany, 1847).

18. Shaw, *Erie Water West,* 344.

19. For illustrations of canal structures see Shaw, *Erie Water West;* Hahn, *The C & O Canal;* Kytle, *Home on the* [Delaware and Raritan] *Canal;* Lee, *Morris Canal;* William J. McKelvey, Jr., *Champlain to Chesapeake: A Canal Era Pictorial Cruise* (Exton, Pa., 1978); Bartholomew and Metz, *Delaware and Lehigh Canals;* Hahn, *The C & O Canal;* and Jack Gieck, *A Photo Album of Ohio's Canal Era, 1825-1913* (Kent, Ohio, 1988).

20. Walter B. Smith, "Wage Rates on the Erie Canal, 1828-1881," *Journal of Economic History* 33 (Sept. 1963): 305; Shaw, *Erie Water West,* 91; Scheiber, *Ohio Canal Era,* 72-73; Fatout, *Indiana Canals,* 84-85; Tobin, "Lowly Muscular Digger," 27, 46, 54. Samuel Forrer reported his wages as a "junior rodman" on the Miami Canal in Ohio in 1823 at $9 a month, which was raised to $12 when he was promoted to "senior rodman" and to $600 a year when he became an assistant engineer in 1824. Forrer noted that "we had few mechanics of any kind in the country at the time" (Forrer, "Autobiography").

21. Way, "Shovel and Shamrock," 498, 506; Scheiber, *Ohio Canal Era,* 73; Taylor, *Transportation Revolution,* 289-91.

22. In 1828, five thousand men were employed on the Pennsylvania canals (Bishop, "State Works of Pennsylvania," 189; Shaw, *Erie Water West,* 121; Sanderlin, *Great National Project,* 78; Tobin, "Lowly Muscular Digger," 42; Fatout, *Indiana Canals,* 84). In addition, two thousand were employed on the Ohio and Erie Canal in 1825, and a thousand men worked on the Blackstone Canal in 1826 (Taylor, *Transportation Revolution,* 289; Powers, "'Invisible Immigrants,'" 111).

23. *Laws of the State of New York,* 1:403; Shaw, *Erie Water West,* 90; Teagarden, "Builders of the Ohio Canal," 95.

24. Quoted in Way, "Shovel and Shamrock," 495.

25. Powers, "'Invisible Immigrants,'" 118-19; Sanderlin, *Great National Project,* 72.

26. Kerby A. Miller, *Emigrants and Exiles: Ireland and the Irish Exodus to North America* (New York, 1985), 20, 98, 193-201, 291-92. The numbers of Irish immigrants coming to Canada and the United States varied with conditions in Ireland and North America, and many Irish came south from Canada. Between 1815 and 1818, twenty thousand Irish a year crossed the Atlantic; nineteen thousand came to North America in 1826; and almost four hundred thousand came between 1828 and 1837.

27. Miller, *Emigrants and Exiles,* 70-71. "Strictly speaking," writes Miller, "Irish country people were not illiterate but preliterate: through the oral medium they transmitted a rich, robust traditional culture."

28. David Grimsted, "Rioting in Its Jacksonian Setting," *American Historical Review,* vol. 77, (April 1972): 390.

29. Way, "Shovel and Shamrock," 490-506, 515-17. See also George Potter, *To the Golden Door: The Story of the Irish in Ireland and America* (Boston, 1960), chap. 32.

30. Stapleton, *Transfer of Industrial Technologies,* 120-21; Shaw, *Erie Water West,* 93.

31. Tobin, "Lowly Muscular Digger," 129; Shaw, *Erie Water West,* 223-24; Sanderlin, *Great National Project,* 93-97; Fickle, "'People' versus 'Progress,'" 320-21. See Charles E. Rosenberg, *The Cholera Years: The United States in 1832, 1849, and 1866* (Chicago, 1962), pt. 1.

32. Quoted in Tobin, "Lowly Muscular Digger," 129-30.

33. Ibid., 132.

34. William J. Rorabaugh, *The Alcoholic Republic: An American Tradition* (New York, 1979), 143-44; Tobin, "Lowly Muscular Digger," 180, 185; Sanderlin, *Great National Project,* 116-23; Way, "Shovel and Shamrock," 490-502.

36. Carl E. Prince, "The Great 'Riot Year': Jacksonian Democracy and Patterns of Violence in 1834," *Journal of the Early Republic,* vol. 5 (Spring 1985): 17-18; Morris, "Andrew Jackson, Strikebreaker," 54-55.

37. Shaw, *Erie Water West,* 236; McIlwraith, "Erie and Great Lakes Canals," 865-67.

38. Earl J. Heydinger, "Lockage Time," *Canal Currents,* no. 67 (Summer 1984): 9; Stilgoe, *Common Landscape of America,* 117.

39. Stapleton, *Transfer of Industrial Technologies,* 70; Calhoun, *American Civil Engineer,* 24-50.

40. Hunter, *Water Power,* chap. 1.

41. Earl J. Heydinger, "Canalling North-South across Pennsylvania," *Canal Currents,* no. 46 (Spring 1979): 10-11; Baer, *Canals and Railroads,* 8-9; Richard D. Brown, *Modernization: The Transformation of American Life, 1600-1865* (New York, 1976), 13, 123-24, 153; Stilgoe, *Common Landscape of America,* 125. See also John F. Stover, "Canals and Turnpikes: America's Early Nineteenth-Century Transportation Network," in Joseph R. Frees and Jacob Judd, eds., *An Emerging Independent American Economy, 1815-1875* (Tarrytown, N.Y., 1980), 60-98.

42. Wabash and Erie Canal tollbook, Allen County-Fort Wayne Historical Society Museum, extracted as "A partial listing of boats passing to or through Fort Wayne on The Wabash and Erie Canal, June-Sept. 1845," *Indiana Waterways* 4 (Spring 1985): 8.

43. "New York and Ohio Line," Broadside (1835), Ohio Historical Society.

44. Baer, *Canals and Railroads,* 9; Scheiber, "Pennsylvania and Ohio Canal," 122-26; Corkan, "Beaver and Lake Erie Canal," 178-90.

45. Harry N. Scheiber, "The Rate-Making Power of the State in the Canal Era: A Case Study," *Political Science Quarterly* 77 (Sept. 1962): 413.

46. Henry B. Stanton, *Random Recollections,* 2d ed. (New York, 1886), 18; David Wilkie, *Sketches of a Summer Trip to New York and the Canadas* (Edinburgh, 1837), quoted in Roger Haydon, ed., *Upstate Travels: British Views of Nineteenth Century New York* (Syracuse, 1982), 145-46.

47. John C. Spencer to James K. Livingston, April 30, 1829; Granger to Weed, Oct. 1832, Thurlow Weed Papers, University of Rochester Library, Rochester, N.Y.

48. *Account of a Journey of Sibyl Tatum with her Parents from N. Jersey to Ohio in 1830* (Independence, Ohio), [n.d.], 8.

49. Stewart Scott, Diary, Aug. 2–Nov. 19, 1826, quoted in Shaw, *Erie Water West,* 206.

50. Hamil Loring to George F. Grimm, Aug. 5, 1846, quoted in ibid., 211.

51. "Sketches from Memory: The Canal-Boat," in Nathaniel Hawthorne, *Tales and Sketches,* ed. Roy Harvey Pearce (New York, 1982), 344-51. The quotations in the following paragraphs are from this source.

52. Charles Dickens, *American Notes for General Circulation* (1842), ed. John S. Whitley and Arnold Goldman (New York, 1972), 191-98.

53. Nevins, *Diary of John Quincy Adams,* 557.

54. Dr. Albert C. Koch, *Journey through a Part of the United States of North America in the Years 1844 to 1846,* trans. and ed. Ernest A. Stadler (Carbondale, Ill., 1972), 33-34, 36-37, 114-15. The quotations in the following paragraph are from the same source.

55. Anne Royall, *The Black Book,* 2 vols. (Washington, D.C., 1828), 1:37.

56. Caroline Gilman, *The Poetry of Travelling in the United States* (New York, 1828), 89-93.

57. Harriet Martineau, *Retrospect of Western Travel,* 2 vols. (New York, 1838), 1:77.

58. Catherine Dickinson to Samuel F. Dickinson, June 15, 1835, reprinted in *Canal Currents,* no. 58 (Spring 1982): 6-7.

59. H.E. Beecher Stowe, "The Canal Boat," *Godey's Lady's Book, and Ladies American Magazine* 23 (Oct. 1841): 167-69.

60. Francis Anne Butler [Kemble], *Journal,* 2 vols. (London, 1835) 2:243-45; Frances Trollope, *Domestic Manners of the Americans* (1832), ed. Donald Smalley (New York, 1949), 369. Trollope also commented on the Chesapeake and Delaware, the Chesapeake and Ohio, and the Morris canals.

61. Emily A. Madden, "Canal Days Letters," *Towpaths* 11, no. 2 (1973): 23.

62. William Cooper Howells, *Recollections of Life in Ohio, from 1813 to 1840* (Cincinnati, 1895), 180.

63. William Dean Howells, *A Boy's Town* (New York, 1890), 239-41.

64. Quoted in Shaw, *Erie Water West,* 211. "We have plenty musical instruments on board," wrote a passenger at Schenectady in 1830.

65. Allan Nevins, ed., *The Diary of Philip Hone, 1828-1851,* 2 vols. (New York, 1927), 1:164-65, 2:804. An account of a round-trip journey by a New Englander on the Chesapeake and Ohio Canal in 1859 is reprinted in Hahn, *The C & O Canal: Pathway to the Nation's Capital* (Metuchen, N.J., 1984), 31-86.

66. Gilman, *Poetry of Travelling,* 93.

67. Fred Gustorf, ed., *The Uncorrupted Heart: Journal and Letters of Frederick Julius Gustorf, 1800-1845* (Columbia, Mo., 1969), 11, 14.

68. "Diary of Mary Pratt, 1831," quoted in Woods, "Crossing Ohio by Boat," 73.

69. *Buffalo Commercial Advertiser,* June 25, 1845.

70. See articles in *Canal Currents,* no. 42 (Spring 1978): 10, 12; no. 48 (Autumn 1979): 11; no. 56 (Autumn 1981): 4.

71. Shaw, *Erie Water West,* 244. Staff of the Canal Museum, Syracuse, N.Y., *A Canalboat Primer on the Canals of New York State* (Syracuse, 1981), 7-9.

72. *Hunt's Merchants' Magazine* 6 (March 1842): 278.

73. Eugene F. Moran, Sr., "The Erie Canal as I Have Known It," *Bottoming Out* 3 (1959): 2-18.

74. Shaw, *Erie Water West,* 244; Scheiber, *Ohio Canal Era,* 261.

75. Quoted in Shaw, *Erie Water West,* 245.

76. Carp, "Erie Canal and Republicanism," 590-94. Lionel D. Wyld, *Low Bridge: Folklore and the Erie Canal* (Syracuse, 1962), 69.

77. Grimsted, "Rioting in Its Jacksonian Setting," 361-97. See also Michael Feldberg, *The Turbulent Era: Riot and Disorder in Jacksonian America* (New York, 1980).

78. Ross Frederick Bagby, "Homelands: The Randolph Slaves Seek Ohio Lands," paper presented at Ohio Academy of History, April 22, 1989.

79. Quoted in Shaw, *Erie Water West,* 231, 293; *Indiana Waterways,* no. 5 (June 1982): 4; James S. Buckingham, *The Eastern and Western States of America,* 3 vols. (London, 1842), 2:346-47.

80. Johnson, *Davis Island Lock and Dam,* 110. Jay Cooke worked as a Philadelphia packet boat runner before entering upon his career in finance (Ellis Paxon Oberholtzer, *Jay Cooke: Financier of the Civil War,* 2 vols. [Philadelphia 1907], 1:40-50.).

81. Paul E. Johnson, *A Shopkeeper's Millennium: Society and Revivals in Rochester, New York, 1815-1837* (New York, 1978), 87.

82. *Niles' Weekly Register* 31 (Oct. 7, 1826): 96.

83. *Buffalo Daily Commercial Advertiser,* July 12, 1836; *Summit Beacon* [April 3, 1850], quoted in Woods, "Crossing Ohio by Packet Boat," 81.

84. Whitney R. Cross, *The Burned-Over District: The Social and Intellectual History of Enthusiastic Religion in Western New York, 1800-1850* (Ithaca, 1950), chap. 4.

85. N. Gordon Thomas, "The Millerite Movement in Ohio," *Ohio History* 81 (Spring 1972): 95-107.

86. Samuel Rezneck, "A Traveling School of Science on the Erie Canal in 1826," *New York History* 40 (July 1959): 255-69.

87. Harlow, *Old Towpaths,* 340, 367. Tom Thumb from P.T. Barnum's circus toured the Pennsylvania canals in the 1830s (*Canal Currents,* no. 44 [Autumn 1978]): 4.

88. Kasson, *Civilizing the Machine,* 41, 172-80.

89. Rosenberg, *Cholera Years,* 121, 137, 228.

90. M. Eaton, *Five Years on the Erie Canal . . .* (Utica, 1845), 11.

91. *Rochester Daily Democrat,* Dec. 22, 1845.

92. *Buffalo Journal,* Sept. 8, 1830.

93. Shaw, *Erie Water West,* 225; Carp, "Erie Canal and Republicanism," 634-36.

94. Rorabaugh, *Alcoholic Republic,* 140, 143-49, Appendix 5.

95. Johnson, *Shopkeeper's Millennium,* 19, 55-61.

96. Shaw, *Erie Water West,* 228. Similar prohibitions were attempted on the Wabash and Erie Canal and on other canals (Tobin, "Lowly Muscular Digger," 95).

97. Nott to Clinton, June 14, 1826, De Witt Clinton Papers, Columbia University Library.

98. Quoted in Carp, "Erie Canal and Republicanism," 600.

99. Shaw, *Erie Water West,* 226-27.

100. Quoted in Carp, "Erie Canal and Republicanism," 605.

101. Harlow, *Old Towpaths,* 362.

102. Howells, *A Boy's Town,* 40-41; *Buffalo Commercial Advertiser,* April 12, 1839.

103. Moritz Busch, *Travels between the Hudson and the Mississippi, 1851-1852,* trans. and ed. Norman H. Binger (Lexington, Ky., 1971), 108, 120.

104. Harriet Martineau, *Society in America,* 2 vols. (New York, 1837), 2:13.

7. The Canal Era in Politics and Economic Development

1. Smith, *Shaping of America,* 3:773.

2. Larson, "A Bridge, a Dam, and a River," 355.

3. Ronald E. Shaw, "Canals in the Early Republic: A Review of Recent Literature," *Journal of the Early Republic* 4 (Summer 1984): passim. See also Robert A. Lively, "The American System: A Review Article," *Business History Review* 29 (March 1955): 81-96; and Carter Goodrich, "Internal Improvements Reconsidered," *Journal of Economic History* 30 (June 1970): 289-311. An early single-volume survey of the Canal Era was Harlow, *Old Towpaths;* another survey is Madeline Sadler Waggoner, *The Long Haul West: The Great Canal Era, 1817-1850* (New York, 1958).

4. Taylor, *Transportation Revolution,* chap. 2. See Harry N. Scheiber and Stephen Salsbury, "Reflections on George Rogers Taylor's *The Transportation Revolution, 1815-1860:* A Twenty-five Year Retrospect," *Business History Review* 51 (Spring 1977): 79-89.

5. See Goodrich, *Government Promotion of American Canals,* and Goodrich, ed., *Canals and American Economic Development.*

6. *Laws of the State of New York,* 1:68. Canals as a product of mixed enterprise are described in canal studies by Carter G. Goodrich, Ralph D. Gray, and Harry N. Scheiber.

7. Scheiber, "The Transportation Revolution and American Law," in *Transportation and the Early Nation,* 12; see also Douglas E. Clanin, "Internal Improvements in National Politics, 1816-1830," in *Transportation and the Early Nation,* 30-60.

8. Wiebe, *Opening of American Society,* 200-203.

9. Calhoun quoted in Hill, *Roads, Rails and Waterways,* 40.

10. Gray, *National Waterway,* 47-49.

11. Quoted in Hill, *Roads, Rails and Waterways,* 32.

12. Ibid., 49-54; Calhoun, *American Civil Engineer,* 38-39; Glyndon G. Van Deusen, *The Jacksonian Era, 1828-1848* (New York, 1959), 54.

13. Hurst, *Law and the Conditions of Freedom,* passim.

14. Scheiber, "The Transportation Revolution and American Law," 22.

15. New-York Corresponding Association, for the Promotion of Internal Improvements, *Public Documents,* Introduction, xlii-xliv; see also Charles G. Haines, *Considerations on the Great Western Canal . . .* (Brooklyn, 1818), 5, 9, 11-12.

16. Colden, *Memoir,* 5.

17. John Seelye, "'Rational Exultation': The Erie Canal Celebration," *Proceedings of the American Antiquarian Society* 94 (Worcester, 1984), part 2, 249.

18. Colden, *Memoir,* 92-93.

19. R.W.B. Lewis, *The American Adam* (Chicago, 1955), 159.

20. Seelye, "'Rational Exultation,'" 263.

21. Goodrich, *Government Promotion of American Canals,* 46; George A. Lipsky, *John Quincy Adams: His Theory and Ideas* (New York, 1950), 148-54; Clanin, "Internal Improvements," 43-44; Wiebe, *Opening of American Society,* 217.

22. Scheiber, *Ohio Canal Era,* (second printing, 1987), xxi.

23. Quoted in ibid., 79.

24. Ibid., 109.

25. Ibid., 198.

26. B.U. Ratchford, *American State Debts* (Durham, N.C., 1941), 95; Reginald C. McGrane, *Foreign Bondholders and American State Debts* (New York, 1935), 9.

27. Leland Hamilton Jenks, *The Migration of British Capital to 1875* (London, 1927), 70-81.

28. Ibid., 74; Taylor, *Transportation Revolution,* 49.

29. Nathan Miller, *The Enterprise of a Free People: Aspects of Economic Development in New York State during the Canal Period, 1792-1838* (Ithaca, N.Y. 1962), 89.

30. Ibid., 110; for a full account of domestic and foreign investment in the Erie Canal, see chaps. 5 and 7.

31. Ibid., chaps. 8-10.

32. Shaw, *Erie Water West,* 310-20.

33. Miller, *Enterprise of a Free People,* 214-15.

34. Shaw, *Erie Water West,* 366-67, 387.

35. Scheiber, "Public Canal Finance and State Banking in Ohio," 119-20, 125; Ernest Ludlow Bogart, *Internal Improvements and State Debt in Ohio* (New York, 1924), 21-22.

36. Scheiber, "Public Canal Finance," 127; Scheiber, *Ohio Canal Era,* 38-51.

37. Scheiber, *Ohio Canal Era,* 151.

38. Ibid., 141. Alfred Kelley was appointed a commissioner of the Canal Fund in 1841. He went to New York desperately seeking loans for Ohio, as he wrote to his wife, "making almost super-human efforts—using my personal influence, pledging my individual responsibility to raise money to pay the interest on the State debt" (quoted in Bates, *Alfred Kelley,* 108-9).

39. Scheiber, *Ohio Canal Era,* 157-58; Ralph W. Hidy, *The House of Baring in American Trade and Finance: English Merchant Bankers at Work, 1763-1861* (Cambridge, Mass., 1949), 290. "Ohio bonds," wrote Hidy, "had the best reception both in England and on the Continent." See also Bogart, *Internal Improvements and State Debt,* 165-78.

40. Rubin, "Imitative Public Improvement," 107; Bishop, "State Works of Pennsylvania," 209-29.

41. Bishop, "State Works of Pennsylvania," 205-12, 213, 215; Zimmerman, "Governments and Transportation Systems," 31-37; Thomas Payne Govan, *Nicholas Biddle: Nationalist and Public Banker* (Chicago, 1959), 285.

42. Hidy, *House of Baring,* 294; Bishop, "State Works of Pennsylvania," 222-23.

43. Hidy, *House of Baring,* 313-21; McGrane, *Foreign Bondholders,* 65-81.

44. Hartz, *Economic Policy and Democratic Thought,* 175; Goodrich, *Government Promotion of American Canals,* 68-73.

45. Hidy, *House of Baring*, 293, 321-22.

46. McGrane, *Foreign Bondholders*, 85-86. In 1829, the company sent Richard Rush to Europe, where he secured an initial loan in Holland and attempted to borrow $10 million for the entire canal (Sanderlin, *Great National Project*, 82).

47. McGrane, *Foreign Bondholders*, 88.

48. Ibid., 91-98; Hidy, *House of Baring*, 322-27.

49. Hidy, *House of Baring*, 323-26.

50. Ibid., 328-30; McGrane, *Foreign Bondholders*, 99-101, 110.

51. Dunaway, *James River and Kanawha Company*, 91.

52. Ibid., 90.

53. Quoted in Jenks, *Migration of British Capital*, 94; Richard L. Morton, "The Virginia State Debt and Internal Improvements, 1820-38," *Journal of Political Economy* 25 (April 1917): 367-73.

54. Dunaway, *James River and Kanawha Company*, 135-57.

55. John A. Munroe, *Louis McLane: Federalist and Jacksonian* (New Brunswick, N.J., 1973), 445-49.

56. Quoted in ibid., 455.

57. Earl J. Heydinger, "Canals 'Hurting' Canals," *Canal Currents*, no. 55 (Summer 1981): 11.

58. Fatout, *Indiana Canals*, 97; McGrane, *Foreign Bondholders*, 132; Ratchford, *American State Debts*, 91; Esarey, *Internal Improvements in Early Indiana*, 124; Kalata, *A Hundred Years, A Hundred Miles*, 359-60.

59. Gray, "Canal Era in Indiana," 120-21.

60. Haeger, *Investment Frontier*, 121, 217.

61. Ibid., 126.

62. Ibid., 221; McGrane, *Foreign Bondholders*, 136-42; Hidy, *House of Baring*, 338-39. The differences in the Butler bills of 1846 and 1847 are described in Esarey, *Internal Improvements in Early Indiana*, 134-43.

63. Haeger, *Investment Frontier*, 205; Putnam, *Illinois and Michigan Canal*, 31.

64. Haeger, *Investment Frontier*, 108; McGrane, *Foreign Bondholders*, 110-11.

65. McGrane, *Foreign Bondholders*, 113.

66. Ibid., 117; Haeger, *Investment Frontier*, 91-93, 204.

67. Hidy, *House of Baring*, 337-38; McGrane, *Foreign Bondholders*, 119, 122-25; Haeger, *Investment Frontier*, 207.

68. Richardson, ed., *Messages and Papers*, 3:122-23.

69. *Morning Courier and New York Enquirer*, May 18, 1852; Samuel B. Ruggles, *Vindication in 1849 of the Canal Policy of the State of New York* (Rochester, 1849), 4. See Steven Edwin Siry, "De Witt Clinton and the American Political Economy: Sectionalism, Politics, and Republican Ideology, 1787-1828" (Ph.D. dissertation, University of Cincinnati, 1986), 275-77.

70. Miller, *Enterprise of a Free People*, passim; De Alva S. Alexander, *Political History of the State of New York*, 4 vols. (New York, 1906-23), 2:63. In New York, as elsewhere, the Hunker or more conservative wing of the Democratic party was more willing to support indebtedness for internal improvements, while the Locofoco or more radical wing was more opposed. See Shaw, *Erie Water West*, chaps. 16-18.

71. Lee Benson, *The Concept of Jacksonian Democracy: New York as a Test Case* (Princeton, 1961), 125-31, 150.

72. Hartz, *Economic Policy and Democratic Thought*, 77, 165.

73. Petrillo, *Anthracite and Slackwater*, 75-84.

74. Scheiber, *Ohio Canal Era,* 158.

75. Ibid., 168.

76. Fatout, *Indiana Canals,* 40-41, 49-50.

77. Haeger, *Investment Frontier,* 220. In New York, however, Butler had been closely allied with the Democratic Albany Regency.

78. Krenkel, *Illinois Internal Improvements,* 181.

79. Ibid., 183, 188.

80. Egerton, *Charles Fenton Mercer,* 256-62.

81. Sanderlin, *Great National Project,* 133-37; Hidy, *House of Baring,* 325, 328-29.

82. Dunaway, *James River and Kanawha Company,* 195.

83. Ibid., 194.

84. Ibid., 199-200.

85. John Bell Rae, "Federal Land Grants in Aid of Canals," *Journal of Economic History* 4 (Nov. 1944): 167-77.

86. Ibid., 168, 170, 174-75; Harry N. Scheiber, "State Policy and the Public Domain: The Ohio Canal Lands," *Journal of Economic History* 25 (March 1965): 86-113; Scheiber, "Land Reform, Speculation, and Governmental Failure: The Administration of Ohio's State Canal Lands, 1836-60," *Prologue: The Journal of the National Archives* 7 (1975): 85-98.

87. Rae, "Federal Land Grants in Aid of Canals," 167-77; Gray, "Canal Era in Indiana," 116-17, 122-23, 125; Krenkel, *Illinois Internal Improvements,* 30-31, 193-94; Neu, "Mineral Lands," 175-91; Dickinson, *To Build a Canal,* chaps. 11-12.

88. Gray, *National Waterway,* chap. 2.

89. Hartz, *Economic Policy and Democratic Thought,* 123-24.

90. William Chazanof, *Joseph Ellicott and the Holland Land Company: The Opening of Western New York* (Syracuse, 1970), 160-76.

91. Taylor, *Transportation Revolution,* 152; Harvey H. Segal, "Cycles of Canal Construction," in Goodrich, ed., *Canals and Development,* 172; Baer, *Canals and Railroads,* Appendix: "Cumulative Canal Mileage 1800-62." See also Poor, *Railroads and Canals,* passim, for mileage figures by state.

92. Segal, "Cycles of Canal Construction," 169-215. Harry N. Scheiber has described a first phase of the transportation revolution when steamboats appeared on the rivers after 1815 and the Erie Canal was opened in 1825, followed by a second phase in the period 1840-51. In the latter phase, new canal routes were added in the West and there was a "sharp decline in freight rates on all the major waterways, both old and new" (*Ohio Canal Era,* chap. 9).

93. Cranmer, "Improvements without Public Funds," 157-66; Gray, "Canal Era in Indiana," 129.

94. Shaw, *Erie Water West,* 299; Peter Temin, *Causal Factors in American Economic Growth in the Nineteenth Century* (London, 1975), 38.

95. Roger L. Ransom, "Social Returns from Public Transport Investment: A Case Study of the Ohio Canal," *Journal of Political Economy* 78 (Sept.-Oct. 1970): 1041-60; Ransom, "Public Canal Investment and the Opening of the Old Northwest," in David C. Klingaman and Richard K. Vedder, eds., *Essays in Nineteenth Century Economic History: The Old Northwest* (Athens, Ohio, 1975), 248-49; Scheiber, *Ohio Canal Era,* 198.

96. Henry V. Poor, *Manual of the Railroads of the United States for 1868-1869* (New York, 1868), 13, cited in Gray, *National Waterway,* 109.

97. Rubin, "An Imitative Improvement," 106-14; Lance E. Davis et al, *American Economic Growth: An Economist's History of the United States* (New York, 1972), 480-81.

98. Harry N. Scheiber, "On the New Economic History—and Its Limitations: A Review Essay," *Agricultural History* 41 (Oct. 1967): 387-88; Taylor, *Transportation Revolution,* 134-36. For broad comparisons, Peter D. McClelland gives freight charges per ton-mile as follows: roads, 10 to 15¢; turnpike roads, 15 to 20¢; canals, 1.5¢; steamboats, 0.5 to 1.5¢; and railroads, 2.5¢ ("Transportation," in Glenn Porter, ed., *Encyclopedia of Economic History: Studies in the Principal Movements and Ideas,* 3 vols. [New York, 1980], 1:325).

99. For an extended summary of this literature see Shaw, "Canals in the Early Republic," 125-26n. Douglas C. North argued in 1961 that canals led the West to specialize in agriculture and the East in manufacturing, and Roger Ransom concluded in 1971 that "the immediate impact of the canals was primarily to stimulate the production of agricultural staples" for export to the East. Conversely, Albert Niemi, Jr., argued that canals encouraged western manufacturing, stifled concentration on agriculture, and promoted regional diversification. See Roger L. Ransom, "Interregional Canals and Economic Specialization in the Ante-Bellum United States," *Explorations in Entrepreneurial History,* 2d ser., 5 (Fall 1967): 12-35. See Douglas C. North, *The Economic Growth of the United States, 1790-1960* (New York, 1961), 102-11; Roger L. Ransom, "Interregional Canals and Economic Specialization in the Ante-Bellum United States," Explorations in Entrepreneurial History, 2d ser., 5 (Fall 1967): 12-35; and Albert Niemi, Jr., "A Further Look at Interregional Canals and Economic Specialization, 1820-1840," *Explorations in Economic History* 7 (Summer 1970), 515.

100. Robert G. Albion, *Rise of New York Port, 1815-1860* (New York, 1939), 164.

101. Blake McKelvey, "The Erie Canal: Mother of Cities," *New-York Historical Quarterly* 11 (July 1949): 1-24; Stuart Blumin, *The Urban Threshold: Growth and Change in a Nineteenth-Century American Community* (Chicago, 1976), 63-65; Cummings, "Pennsylvania" 260-73; Harry N. Scheiber, "Urban Rivalry and Internal Improvements in the Old Northwest, 1820-1860," *Ohio History* 71 (Oct. 1962): 228-39; Scheiber, "Ohio's Transportation Revolution—Urban Dimensions, 1803-1870," in John Wunder, ed., *Toward an Urban Ohio* (Columbus, 1977), 14-17; Edwin Maldonado, "Urban Growth during the Canal Era: The Case of Indiana," *Indiana Social Studies Quarterly* 31 (Winter 1978-79): 20-37.

102. Vivienne Dawn Maddox, "The Effect of the Erie Canal on Building and Planning in Syracuse, Palmyra, Rochester, and Lockport, New York" (Ph.D. dissertation, Cornell University, 1976). See also Richard L. Ehrlich, "The Development of Manufacturing in Selected Counties in the Erie Canal Corridor, 1815-1860" (Ph.D. dissertation, State University of New York at Buffalo, 1972).

103. Roberta Balstad Miller, *City and Hinterland: A Case Study of Urban Growth and Regional Development* (Westport, Conn., 1979); Preston, "Market and Mill Town"; Diane Lindstrom, *Economic Development in the Philadelphia Region, 1810-1850* (New York, 1978), 117-19, 185. For a contrasting view that sees canals as less responsible for growth see Simeon J. Crowther, "Urban Growth in the Mid-Atlantic States, 1785-1850," *Journal of Economic History* 36 (Sept. 1976): 641.

104. Rubin, "Canal or Railroad?" 27, 38.

105. Kirkland, *Men, Cities and Transportation,* 1:chaps. 4-15.

106. Albert Fishlow, *American Railroads and the Transformation of the Ante-Bellum Economy* (Cambridge, Mass., 1965), chaps. 2, 7.

107. Whitford, *History of the Canal System of New York,* 1:911. George Rogers Taylor wrote that "in terms of tons of western produce moved eastward to tidewater, the Erie Canal was still the predominant agency in 1860" when the tonnage carried was 1,896,975 (*Transportation Revolution,* 167).

Bibliographical Essay

Early Transportation Studies

Early transportation studies include Caroline E. MacGill *et al.*, *History of Transportation in the United States before 1860* (Washington 1917); Seymour Dunbar, *History of Travel in America,* 4 vols. (Indianapolis 1915); A.B. Hulbert, *Historic Highways of America,* 16 vols. (Cleveland 1902-1905); Ulrich B. Phillips, *A History of Transportation in the Eastern Cotton Belt to 1860* (New York 1908); and Noble E. Whitford, *History of the Canal System of the State of New York,* 2 vols. (Albany 1906). Canal monographs include George Washington Ward, *The Early Development of the Chesapeake and Ohio Canal Project* (Johns Hopkins University Studies in Historical and Political Science (Baltimore 1899); Chester L. Jones, *The Economic History of the Anthracite-Tidewater Canals* (University of Pennsylvania Publications in Political Economy and Public Law, Philadelphia 1908); Wayland Fuller Dunaway, *History of the James River and Kanawha Company* (Columbia University Studies in History, Economics and Public Law, New York 1922); T.B. Klein, *The Canals of Pennsylvania and the System of Internal Improvements* (Harrisburg 1901); Avard Longley Bishop, *The State Works of Pennsylvania* (Transactions, Connecticut Academy of Arts and Sciences, New Haven 1908); James William Putnam, *The Illinois and Michigan Canal: A Study in Economic History* (Chicago Historical Society Collections, Chicago 1918); C.P. McClelland and C.C. Huntington, *History of the Ohio Canals* (Columbus, 1905); Logan Esarey, *Internal Improvements in Early Indiana* (Indiana Historical Society *Publications,* Indianapolis 1912); and Elbert J. Benton, *The Wabash Trade Route in the Development of the Old Northwest* (Johns Hopkins University Studies in Historical and Political Science, Baltimore 1903).

Later Comprehensive Canal Studies

Alvin F. Harlow, *Old Towpaths: The Story of the American Canal Era* (New York and London 1926) was the first comprehensive survey of the Canal Era and it provides a point of departure for more recent canal histories. Twenty-five years later George Rogers Taylor's seminal volume, *The Transportation Revolution,*

1815-1860 (New York 1951) placed American canals in an entirely new conceptual framework, that of public economic policy in an emerging national economy. Carter Goodrich followed with *Government Promotion of American Canals and Railroads, 1800-1890* (New York 1960), which emphasized the canals that broke the Appalachian barrier, and he advanced the thesis of the activist state contributing to economic development through decentralized government and mixed enterprise. Goodrich also collaborated with Julius Rubin, H. Jerome Cranmer, and Harvey H. Segal in the volume of regional and developmental essays, *Canals and American Economic Development,* ed. Carter Goodrich (New York 1961). These two volumes, together with the work of George Rogers Taylor, provide a basic political and economic analysis of American canals.

More popular surveys of American canals are Madeline Sadler Waggoner, *The Long Haul West: The Great Canal Era, 1817-1850* (New York 1958), and Harry Sinclair Drago, *Canal Days in America: The History and Romance of Old Towpaths and Waterways* (New York 1972).

Canals of New York and New England

The essential economic study of the Erie Canal is Nathan Miller, *The Enterprise of a Free People: Aspects of Economic Development in New York State during the Canal Period, 1792-1838* (Ithaca 1962), which emphasizes the contribution of the Canal Fund to the economic development of New York. Political and social aspects of the history of the Erie Canal are described in Ronald E. Shaw, *Erie Water West: A History of the Erie Canal 1792-1854* (Lexington, Ky., 1966). Julius Rubin undertook a fresh examination of the decision to build the Erie Canal in his chapter, "An Innovating Public Improvement: The Erie Canal," in Goodrich, ed., *Canals and American Economic Development.* The relationships between the Erie Canal and republicanism are studied in depth in Roger E. Carp, "The Erie Canal and the Liberal Challenge to Classical Republicanism, 1785-1850," (Ph.D. diss., Univ. of North Carolina at Chapel Hill, 1986).

Several local or regional studies relate the Erie Canal to particular parts of New York State. Among them are Robert G. Albion, *The Rise of New York Port, 1815-1860* (New York 1939); David M. Ellis, *Landlords and Farmers in the Hudson-Mohawk Region, 1790-1850* (Ithaca 1946); Neal Adams McNall, *An Agricultural History of the Genesee Valley, 1790-1860* (Philadelphia 1952); Paul D. Evans, *The Holland Land Company* (Buffalo 1924); William Chazanof, *Joseph Ellicott and the Holland Land Company: The Opening of Western New York* (Syracuse 1970); Whitney R. Cross, *The Burned-Over District* (Ithaca 1950); and Paul E. Johnson, *A Shopkeeper's Millennium: Society and Revivals in Rochester, New York, 1815-1837* (New York 1978). The Erie Canal has received more popular treatment in Harvey Chalmers II, *The Birth of the Erie Canal* (New York 1960) and Lionel D. Wyld, *Low Bridge! Folklore and the Erie Canal* (Syracuse 1962).

The basic scholarly source for New England canals is Edward C. Kirkland, *Men, Cities, and Transportation: A Study in New England History, 1820-1900*, 2 vols. (Cambridge, Mass. 1948). Louis C. Hunter has written a massive study of hydraulic engineering in New England, which concentrates on the short canals that powered the Lowell mills, *Water Power in the Century of the Steam Engine*, vol. I of *A History of Industrial Power in the United States, 1780-1930* (Charlottesville, Va. 1979). Christopher Roberts, *The Middlesex Canal 1793-1860* (Cambridge, Mass. 1938), is a thorough study of a pioneer canal in New England. For the New Haven and Northampton Canal see Charles R. Harte, *Connecticut's Canals* (Hartford, Conn. 1938), and James M. Camposeo, "The History of the Canal System Between New Haven and Northampton [1822-1847], *Historical Journal of Western Massachusetts* 6 (1977): 37-53. There is much detail on the Blackstone Canal in James B. Hedges, *The Browns of Providence Plantations*, 2 vols. (Providence, R.I. 1968)

Mid-Atlantic Canals: Pennsylvania and New Jersey

A brief but very useful study of the mid-Atlantic canals is Christopher T. Baer, *Canals and Railroads of the Mid-Atlantic States, 1800-1860* (Wilmington, Del. 1981), which is distinguished for its statistical data, its bibliography, and its large-scale maps showing stages of canal construction. The literature on Pennsylvania canals is extensive, but a definitive history remains to be written. Louis Hartz broke new ground in his study of governmental economic policy, *Economic Policy and Democratic Thought: Pennsylvania, 1776-1860* (Cambridge, Mass. 1948). For Julius Rubin the Pennsylvania Mainline system emerged as a mistaken decision in the choice of canal or railroad as a response to the Erie Canal in New York. His thesis is developed most fully in *Canal or Railroad? Imitation and Innovation in the Response to the Erie Canal in Philadelphia, Baltimore, and Boston* (American Philosophical Society Transactions, Philadelphia 1961), and is also presented in his chapter, "An Imitative Public Improvement: The Pennsylvania Mainline," in Goodrich, ed., *Canals and American Economic Development*. Robert McCullough and Walter Leuba, *The Pennsylvania Main Line Canal* (York, Pa. 1973), is a clearly written, informative account of the development and operation of the Mainline system. All of the divisions of the Mainline system, as well as the private company canals, are included in William H. Shank, *The Amazing Pennsylvania Canals*, 150th Anniversary edition (York, Pa. 1981), and the appendix has useful tables of canal data.

Manville B. Wakefield has described the link by canal from Pennsylvania through New York to the Hudson River in *Coal Boats to Tidewater; the Story of the Delaware and Hudson Canal* (South Fallsburg, N.Y. 1965). One of the most detailed accounts of canal building in the upper Susquehanna coal region is F. Charles Petrillo, *Anthracite and Slackwater: The North Branch Canal* (Easton, Pa. 1986). Turning to southeastern Pennsylvania, Edward J. Gibbons has written the most detailed account of the creation of the Schuylkill Navigation in "The Building of the Schuylkill Navigation System, 1815-1828," *Pennsylvania History,* 57 (1990): 13-43, and the history of this system is carried forward in Walter S. Sanderlin,"The Expanding Horizons of the Schuylkill Navigation Company, 1815-1870," *Pennsylvania History* 36 (1969): 174-91. An article by James W. Livingood, "The Canalization of the Lower Susquehanna," *Pennsylvania History* 8 (1941): 131-49, preceded the publication of his book-length study on the trade rivalry between Baltimore and Philadelphia. A second account of these canals is Gerald Smeltzer, *Canals Along the Lower Susquehanna* (York, Pa. 1963). For the Delaware Division Canal along the lower Delaware River see C.P. Yoder, *Delaware Canal Journal: A Definitive History of the Canal and the River Valley through Which It Flows* (Bethlehem, Pa. 1972). This canal is included in the striking pictorial history compiled by Ann Bartholomew and researched by Lance E. Metz, *Delaware and Lehigh Canals* (Easton, Pa. 1989). E.J. Hartman has recognized the contributions of Josiah White and Erskine Hazard to the Lehigh Canal in "Josiah White and the Lehigh Canal," *Pennsylvania History,* 7 (1940): 225-35, as has Donald Sayenga in "The Untryed Business: An Appreciation of White and Hazard," *Proceedings of the Canal History and Technology Symposium* 2 (1983): 105-28. A broader account of the Lehigh Canal is Thomas Dinkelacker, "The Construction of the Lehigh Canal and the Early Development of the Lehigh Valley Region," *Proceedings and Collections of the Wyoming Historical and Genealogical Society* 24 (1984): 45-64. For the history of the Union Canal see Richard N. Pawling, "Geographic Influences upon the Development and Decline of the Union Canal," *Proceedings of the Canal History and Technology Symposium* 2 (1983): 69-85.

The canal along the west branch of the upper Susquehanna River is described in Sidney Davis, "The West Branch Canal," *Proceedings of the Northumberland County Historical Society* 26 (1974): 28-43, and a useful source for the canal connection to Lake Erie in Northwestern Pennsylvania is Lloyd A.M. Corkan, "The Beaver and Lake Erie Canal," *The Western Pennsylvania Historical Magazine* 17 (1934): 175-88. Harry N. Scheiber has examined the interconnections of the canals in northwestern Pennsylvania with the Ohio canals in "The Pennsylvania & Ohio Canal: Transport Innovation, Mixed Enterprise, and Urban Rivalry, 1825-1861," *The Old Northwest* 6 (1980): 105-35. The sharp rivalry between two engineers deeply involved with the Pennsylvania canals is recounted in Donald Sayenga, *Ellet and Roebling* (York, Pa. 1983).

The transport of Pennsylvania coal to eastern markets dominates the histories of the New Jersey canals. Barbara N. Kalata, *A Hundred Years, A Hundred Miles: New Jersey's Morris Canal* (Morristown, N.J. 1983) is a comprehensive history of that canal through northern New Jersey, organized chronologically, which emphasizes the activities of the Morris Canal and Banking Company, the coal trade, and the stimulus of the canal to railroad development. Still useful is the account by the early historian of New Jersey transportation, Wheaton J. Lane, "The Morris Canal," *Proceedings of the New Jersey Historical Society* 55 (1937): 215-63. James Lee, *The Morris Canal: A Photographic History* (Easton, Pa. 1979), shows the inclined planes in use on this canal. Two books describe the Delaware and Raritan Canal in southern New Jersey: James and Margaret Cawley, *Along the Delaware and Raritan Canal* (Rutherford, N.J. 1970), and Elizabeth G.C. Menzies, *Passage between Rivers: A Portfolio of Photographs with a History of the Delaware and Raritan Canal* (New Brunswick, N.J. 1976). H. Jerome Cranmer has used the New Jersey Canals to designate developmental and exploitative canals in his chapter, "Improvements Without Public Funds: The New Jersey Canals" in Goodrich, ed., *Canals and American Economic Development.*

Chesapeake and Southern Canals

The basic source for the Chesapeake and Ohio Canal is Walter S. Sanderlin, *The Great National Project: A History of the Chesapeake and Ohio Canal* (Johns Hopkins University Studies in Historical and Political Science, Baltimore 1946). Another account is *The Chesapeake & Ohio Canal: Pathway to the Nation's Capital* (Metuchen, N.J. 1984) by Thomas F. Hahn, who has also published a *Towpath Guide to the Chesapeake & Ohio Canal: Georgetown Tidelock to Cumberland,* combined edition (Shepherdstown, W.Va. 1982). An engaging informal history of the Chesapeake and Ohio Canal is Elizabeth Kytle, *Home on the Canal* (Cabin John, Md. 1983).

The definitive history of The Chesapeake and Delaware Canal is Ralph D. Gray, *The National Waterway: A History of the Chesapeake and Delaware Canal, 1769-1965* (Urbana and London 1967), which has been published in a new edition in 1989 with an added chapter for the period 1965-1985. The Chesapeake and Delaware Canal was central to the trade rivalry between Philadelphia and Baltimore, a topic examined in a classic study by James Weston Livingood, *The Philadelphia-Baltimore Trade Rivalry 1780-1860* (Harrisburg, Pa. 1947). Gray traced the efforts of Philadelphia merchants and planners to secure a Chesapeake and Delaware Canal in "Philadelphia and Chesapeake and Delaware Canal 1769-1823," *The Pennsylvania Magazine of History and Biography* 84 (1960): 401-23.

In Virginia, the James River and Kanawha Canal awaits a comprehensive study. For canals connecting Chesapeake Bay and Albemarle Sound see two studies by Alexander Crosby Brown, *The Dismal Swamp Canal,* rev. ed. (Chesapeake, Va. 1970), and *Juniper Waterway: A History of the Albemarle and Chesapeake Canal* (Charlottesville, Va. 1981). The Dismal Swamp Canal and the Albemarle and Chesapeake Canal are also described in Clifford R. Hinshaw, Jr., "North Carolina Canals before 1860," *North Carolina Historical Review* 25 (1948): 1-56. A chapter in Henry Savage, Jr., *River of the Carolinas: The Santee* (Chapel Hill, N.C. 1956) places the Santee and Cooper Canal in its South Carolina lowland setting. Milton S. Heath has contributed an outstanding political and economic study which includes the Georgia canals, *Constructive Liberalism: The Role of the State in Economic Development in Georgia to 1860* (Cambridge, Mass. 1954). For an instance of state assistance to a private canal company in the bayou country of south central Louisiana, see Thomas A. Becnel, *The Barrow Family and the Barataria and Lafourche Canal: The Transportation Revolution in Louisiana, 1829-1925* (Baton Rouge 1989).

Canals in the Old Northwest

A survey account of western canals is Ronald E. Shaw, "The Canal Era in the Old Northwest," in *Transportation and the Early Nation* (Indianapolis 1982). The definitive history of the Ohio canals is Harry N. Scheiber, *Ohio Canal Era: A Case Study of Government and the Economy, 1820-1861* (Athens, Ohio 1969), reprinted with a new preface in 1987. Jack Gieck has drawn together hundreds of photographs showing parts of the Ohio canals operating into the twentieth century in *A Photo Album of Ohio's Canal Era, 1825-1913* (Kent, Ohio 1988). Paul Fatout, *Indiana Canals* (West Lafayette, Ind. 1972), reprinted in 1989, offers a lively critical account. The essay by Ralph D. Gray, "The Canal Era in Indiana," in *Transportation and the Early Nation* stresses the relation of federal land grants to the Wabash and Erie Canal and its contribution to economic development. John H. Krenkel, *Illinois Internal Improvements 1818-1848* (Cedar Rapids, Iowa 1958), gives greatest attention to political and administrative aspects of the history of the Illinois and Michigan Canal. A brief account is John M. Lamb, "Early Days on the Illinois & Michigan Canal," *Chicago History* 3 (1974-75): 168-76.

Two authors have written outstanding studies on the Sault Canal, or St. Mary's Falls Canal, in Michigan. John Dickinson, *To Build a Canal; Sault Ste. Marie, 1853-1854 and After* (Columbus 1981), is a comprehensive history that stresses the role of John W. Brooks in completing the canal. Irene D. Neu explored the relationship of the Sault Canal to the federal land grant of 1852 and the company that built the canal and sold the mineral lands. See Irene D. Neu, "The Building of the Sault Canal, 1852-1855," *Mississippi Valley Historical Review,* 40 (1953): 25-46, and "The Mineral Lands of the St. Mary's Falls Ship

Canal Company," in *The Frontier in American Development: Essays in Honor of Paul Wallace Gates*, ed. David M. Ellis (Ithaca, N.Y. 1969): 162-91.

An account of the Louisville and Portland Canal, a canal at the Falls of the Ohio comparable to the Sault Canal, is Paul B. Trescott, "The Louisville and Portland Canal Company, 1825-1874," *Mississippi Valley Historical Review,* 44 (1958): 689-708. For a comprehensive examination of grain exports from the canals of the Old Northwest see John W. Clark, *The Grain Trade in the Old Northwest* (Urbana, Ill. 1966).

Special Topics in Canal History

Study of American canal engineering should begin with Darwin H. Stapleton, *The Transfer of Early Industrial Technologies to America* (Philadelphia 1987), which includes the contributions of William Weston, Benjamin H. Latrobe, and Moncure Robinson. Daniel H. Calhoun, *The American Civil Engineer: Origins and Conflict* (Cambridge, Mass. 1960), describes the work of Loammi Baldwin the younger, John L. Sullivan and Benjamin Wright, and notes the transition from the individualistic engineer to the organization engineer. The contribution of the army engineers to American canal construction is included in Forest G. Hill, *Roads, Rails, & Waterways: The Army Engineers and Early Transportation* (Norman, Okla. 1957). William H. Shank, *Towpaths to Tugboats: A History of American Canal Engineering* (York, Pa. 1982), lists the engineers who served on American canals and provides useful sketches on the most prominent engineers.

John F. Stover has described the transition from turnpikes to canals in an essay, "Canals and Turnpikes: America's Early-Nineteenth-Century Transportation Network," in *An Emerging Independent American Economy 1815-1875,"* ed. Joseph R. Frese, S.J. and Jacob Judd (Tarrytown, N.Y. 1980). For the broader context of canal technology see Lewis Mumford, *Technics and Civilization* (New York 1934); John R. Stilgoe, *Common Landscape of America, 1580-1845* (New Haven, 1982); and Elting E. Morison, *From Know-How to Nowhere: The Development of American Technology* (New York 1974). Morison selects John B. Jervis as almost the embodiment of American engineering in his time. The experiences of this engineer are recorded in Neal FitzSimmons, ed., *The Reminiscences of John B. Jervis, Engineer of the Old Croton* (Syracuse 1971).

Two essays that present the constitutional and political issues related to American canals are Harry N. Scheiber, "The Transportation Revolution and American Law: Constitutionalism and Public Policy," and Douglas E. Clanin, "Internal Improvements in National Politics, 1816-1830," in *Transportation and the Early Nation.* On the Gallatin Plan of 1808 see Carter Goodrich, "The Gallatin Plan after One Hundred and Fifty Years," *American Philosophical*

Society Proceedings 102 (1958): 436-41; and for its implementation see Lee W. Formwalt, "Benjamin Henry Latrobe and the Revival of the Gallatin Plan of 1808," *Pennsylvania History* 48 (1981): 99-128. Two articles that center on internal improvements and Jeffersonian policy are John L. Larson, " 'Bind the Republic Together': The National Union and the Struggle for a System of Internal Improvements," *The Journal of American History* 74 (1987): 363-87, and Joseph H. Harrison, Jr., "*Sic Et Non:* Thomas Jefferson and Internal Improvement," *Journal of the Early Republic* 7 (1987): 335-49. Harrison's article draws upon his unpublished Ph.D. dissertation, "The Internal Improvement Issue in the Politics of the Union 1783-1825," (University of Virginia, 1954), which remains one of the most penetrating studies on the subject.

For the land grants that made some canals virtually land-grant waterways see John Bell Rae, "Federal Land Grants in Aid of Canals," *Journal of Economic History* 4 (1944): 167-77. Harry N. Scheiber has studied land grants in Ohio in "State Policy and the Public Domain: The Ohio Canal Lands," *Journal of Economic History* 25 (1965): 86-113, and "Land Reform, Speculation, and Government Failure: The Administration of Ohio's State Canal Lands, 1836-60," *Prologue: The Journal of the National Archives* 7 (1975): 85-98.

Sources for public and private investment and state debts for canals include the following: John Denis Haeger, *The Investment Frontier: New York Businessmen and the Economic Development of the Old Northwest* (Albany, N.Y. 1981); B.U. Ratchford, *American State Debts* (Durham, N.C. 1941); Reginald C. McGrane, *Foreign Bondholders and American State Debts* (New York 1935); Ralph W. Hidy, *The House of Baring in American Trade and Finance: English Merchant Bankers at Work 1763-1861* (Cambridge, Mass. 1949); and Harvey H. Segal, "Cycles of Canal Construction," in Goodrich, ed., *Canals and American Economic Development.*

Quantitative economic studies have assessed the profitability of American canals and their contribution to economic development. For such analysis see the following studies by Roger L. Ransom and others; Ransom, "Canals and Development: A Discussion of the Issues," *American Economic Review* 54 (1964): 365-76; Ransom, "Interregional Canals and Economic Specialization in the Ante-Bellum United States," *Explorations in Entrepreneurial History,* 2nd ser., 5 (1967): 12-35; Ransom, "Social Returns from Public Transport Investment: A Case Study of the Ohio Canal," *Journal of Political Economy,* 78 (1970): 1041-60; Ransom, "Public Canal Investment and the Opening of the Old Northwest," in David C. Klingaman and Richard K. Vedder, eds., *Essays in Nineteenth Century Economic History: The Old Northwest* (Athens, Ohio 1975); Albert Niemi, Jr., "A Further Look at Interregional Canals and Economic Specialization, 1820-1840," *Explorations in Economic History* 2nd ser. 7 (1970): 499-520; Stanley Lebergott, "United States Transport Advance and Externalities," *Journal of Economic History,* 26 (1966): 437-61; and Harvey H.

Segal, "Canals and Economic Development," in Goodrich, ed., *Canals and American Economic Development.* A benchmark article summarizing canal histories before 1970 is Carter Goodrich, "Internal Improvements Reconsidered," *Journal of Economic History* 30 (1970): 289-311.

An important dimension of canal studies has been the effects of American canals on urban growth. Blake McKelvey pioneered in this field with his article, "The Erie Canal, Mother of Cities," *The New-York Historical Quarterly* 35 (1951): 55-71, and his studies of Rochester, New York. Roberta Balstad Miller used the approach of Eric C. Lampard in her study of Syracuse and its Onondaga County hinterland, *City and Hinterland: A Case Study of Urban Growth and Regional Development* (Westport, Conn. 1979). Stuart Blumin described the process by which Kingston, New York, near the terminus of the Delaware and Hudson Canal, crossed the urban threshold after the canal was completed in *The Urban Threshold: Growth and Change in a Nineteenth-Century American Community* (Chicago 1976). In Pennsylvania Diane Lindstrom calculated a decisive contribution to the growth of Philadelphia and its hinterland from canals in her study, *Economic Development in the Philadelphia Region 1810-1850* (New York 1978). The flavor of life in inland Pennsylvania canal towns and cities is conveyed in Hubertis M. Cummings, "Pennsylvania: Network of Canal Ports," *Pennsylvania History,* 21 (1954): 260-73. Similar studies describe urban growth stimulated by canals in Ohio and Indiana. Harry N. Scheiber traced urban development at strategic points along Ohio canals in his chapter, "Ohio's Transportation Revolution—Urban Dimensions, 1803-1870," in John Wunder, ed., *Toward an Urban Ohio* (Columbus, 1977), and Edwin Maldonado charted similar cities in "Urban Growth During the Canal Era: The Case of Indiana," *Indiana Social Studies Quarterly* 31 (1978-1979): 20-39. The outstanding study by Charles R. Poinsatte, *Fort Wayne During the Canal Era 1828-1855* (Indianapolis 1969), is a major contribution to midwestern urban history as it describes the rise of Fort Wayne from an Indian trading center to a canal port on the Wabash and Erie Canal.

Reflecting the new social history, new scholarship on American canals has turned to study of ethnicity and canal laborers. Catherine Tobin has written on "Irish Labor on American Canals," *Proceedings of the Canal History and Technology Symposium* 9 (1990): 2-35. Studying labor on an eastern canal, Peter Way centers on the Irish secret society, labor violence, and class struggle in "Shovel and Shamrock: Irish Workers and Labor Violence in the Digging of the Chesapeake and Ohio Canal," *Labor History* 30 (1989): 489-517.

Index